Communications
in Computer and Information Science 741

Commenced Publication in 2007
Founding and Former Series Editors:
Alfredo Cuzzocrea, Orhun Kara, Dominik Ślęzak, and Xiaokang Yang

More information about this series at http://www.springer.com/series/7899

Cédric Grueau · Robert Laurini
Jorge Gustavo Rocha (Eds.)

Geographical Information Systems Theory, Applications and Management

Second International Conference, GISTAM 2016
Rome, Italy, April 26–27, 2016
Revised Selected Papers

 Springer

Editors
Cédric Grueau
Polytechnic Institute of Setúbal/IPS
Setúbal
Portugal

Robert Laurini
INSA Lyon
University of Lyon
Villeurbanne Cedex
France

Jorge Gustavo Rocha
Departamento de Informatica
Universidade do Minho
Braga
Portugal

ISSN 1865-0929 ISSN 1865-0937 (electronic)
Communications in Computer and Information Science
ISBN 978-3-319-62617-8 ISBN 978-3-319-62618-5 (eBook)
DOI 10.1007/978-3-319-62618-5

Library of Congress Control Number: 2017945724

Printed on acid-free paper

This Springer imprint is published by Springer Nature
The registered company is Springer International Publishing AG
The registered company address is: Gewerbestrasse 11, 6330 Cham, Switzerland

Preface

The present book includes extended and revised versions of a set of selected papers from the Second International Conference on Geographical Information Systems Theory, Applications, and Management (GISTAM 2016), held in Rome, Italy, during 26–27.

GISTAM 2016 received 33 paper submissions from 25 countries, of which 30% are included in this book. The papers were selected by the event chairs and their selection is based on a number of criteria that include the classifications and comments provided by the Program Committee members, the session chairs' assessment, and also the program chairs' global view of all papers included in the technical program. The authors of selected papers were then invited to submit a revised and extended version of their papers having at least 30% innovative material.

GISTAM aims at creating a meeting point of researchers and practitioners to address new challenges in geospatial data sensing, observation, representation, processing, visualization, sharing, and managing, in all aspects concerning both information communication and technologies (ICT) as well as management information systems and knowledge-based systems. The conference welcomes original papers of either practical or theoretical nature, presenting research or applications, of specialized or interdisciplinary nature, addressing any aspect of geographic information systems and technologies.

The papers selected to be included in this book contribute to the understanding of relevant trends of current research on geographical information systems theory, applications and management, including: urban and regional planning, water information systems, geospatial information and technologies, spatio-temporal database management, decision support systems, energy information systems, and GPS and location detection.

We would like to thank all the authors for their contributions and also the reviewers who helped ensure the quality of this publication.

February 2017

Cédric Grueau
Robert Laurini
Jorge Gustavo Rocha

Organization

Conference Chair

Cédric Grueau   Polytechnic Institute of Setúbal/IPS, Portugal

Program Chair

Jorge Gustavo Rocha   University of Minho, Portugal

Honorary Chair

Robert Laurini   INSA, University of Lyon, France

Program Committee

Masatoshi Arikawa	The University of Tokyo, Japan
Pedro Arnau	Universitat Politècnica de Catalunya, Spain
Arnaud banos	CNRS, France
Piero Boccardo	Politecnico di Torino, Italy
Arnold K. Bregt	Wageningen University and Research Centre, The Netherlands
Alexander Brenning	Friedrich Schiller University, Germany
Rex G. Cammack	University of Nebraska in Omaha, USA
Filiberto Chiabrando	Politecnico di Torino, DIATI, Italy
Jordi Corbera	Geological and Cartographic Institute of Catalonia, Spain
Suzana Dragicevic	Simon Fraser University, Canada
Qingyun Du	Wuhan University, China
Arianna D'Ulizia	IRPPS, CNR, Italy
Ana Paula Falcão	Instituto Superior Técnico, Portugal
Ana Fonseca	Laboratório Nacional de Engenharia Civil (LNEC), Portugal
Luis Gomez-Chova	Universitat de València, Spain
Célia Gouveia	Universidade de Lisboa, Portugal
Cédric Grueau	Polytechnic Institute of Setúbal/IPS, Portugal
Hans W. Guesgen	Massey University, New Zealand
Bob Haining	University of Cambridge, UK
Stephen Hirtle	University of Pittsburgh, USA
Haosheng Huang	University of Zurich, Switzerland
Andrew Hudson-Smith	University College London, UK
Karsten Jacobsen	Leibniz Universität Hannover, Germany
Simon Jirka	52° North, Germany

Stéphane Joost	Ecole Polytechnique Fédérale de Lausanne (EPFL), Switzerland
Harry D. Kambezidis	National Observatory of Athens, Greece
Robert Laurini	Knowledge Systems Institute, USA
Jun Li	Sun Yat-Sen University, China
Christophe Lienert	Canton of Aargau, Department of Construction, Traffic and Environment, Switzerland
Andrea Lingua	Politecnico di Torino, Italy
Gavin McArdle	University College Dublin, Ireland
Richard Milton	University College London, UK
Lan Mu	University of Georgia, USA
Peter Nijkamp	Free University of Amsterdam, The Netherlands
Scott Orford	Cardiff University, UK
Matthew Owen	Cathie Associates/University College London, UK
Dimitris Potoglou	Cardiff University, UK
Guoyu Ren	National Climate Center, China
Mathieu Roche	Cirad, France
Armanda Rodrigues	Universidade Nova de Lisboa, Portugal
Markus Schneider	University of Florida, USA
Yosio Edemir Shimabukuro	Instituto Nacional de Pesquisas Espaciais, Brazil
Jantien Stoter	Delft University of Technology, The Netherlands
H.J.P. Timmermans	Technical University, The Netherlands
Fabio Giulio Tonolo	ITHACA, Information Technology for Humanitarian Assistance, Cooperation and Action, Politecnico di Torino, Italy
Michael Vassilakopoulos	University of Thessaly, Greece
Miguel A. Vengazones	GIPSA-Lab, CNRS, France
Jan Oliver Wallgrün	The Pennsylvania State University, USA
Ouri Wolfson	University of Illinois at Chicago, USA
May Yuan	University of of Texas at Dallas, USA
Xiaojun Yuan	University at Albany, SUNY, USA

Invited Speakers

Roberto Lattuada	myHealthbox, Italy
Wolfgang Kainz	University of Vienna, Austria
Barbara Koch	University of Freiburg, Germany

Contents

Fast Displacements Detection Techniques Considering Mass-Market GPS L1 Receivers

Paolo Dabove$^{(\boxtimes)}$ and Ambrogio Maria Manzino

Department of Environment, Land and Infrastructure Engineering,
Politecnico di Torino, Corso Duca Degli Abruzzi 24, Turin, Italy
{paolo.dabove, ambrogio.manzino}@polito.it

Abstract. Fast displacements detection in real-time is a very high challenge due to the necessity to preserve buildings, infrastructures and the human life. In this paper this problem is addressed using some statistical techniques and a GPS mass-market receiver in real-time. Very often, most of landslides monitoring and deformation analysis are carried out by using traditional topographic instruments (e.g. total stations) or satellite techniques such as GNSS geodetic receivers, and many experiments were carried out considering these types of instruments. In this context it is fundamental to detect whether or not deformation exists, in order to predict future displacement. Filtering means are essential to process the diverse noisy measurements (especially if low cost sensors are considered) and estimate the parameters of interest. In this paper some results obtained considering mass-market GPS receivers coupled with statistical techniques are considered in order to understand if there are any displacements from a statistical point of view in real time. Instruments considered, tests, algorithms and results are herein reported.

Keywords: Fast displacements detection · GNSS NRTK positioning · Mass-Market receivers · Statistical analysis · Accuracy

1 Introduction

Deformation monitoring is the act of ordinary and continuous observation of such variations that are referred to as "deformation" [9, 51].

Considering the types of network, deformation survey techniques are classified as Absolute Deformation Monitoring (some reference points located in the area surrounding the object of interest, i.e. dam, bridge, etc.) and Relative Deformation Monitoring (where the reference points are located in the structure and both the object and reference points are subject to displacement) where the main goal is to estimate the relative displacement between two or more points [1]. Methods of deformation monitoring have changed considerably in principle over the past few decades as newer data sources have come to be used. In sparsely vegetated terrain, landslides are routinely detected and mapped by a combination of the interpretation of airphotos or multi-spectral digital imagery and selective ground verification [5]. However, it is quite difficult to use these methods in rugged terrain covered with dense vegetation. Also,

© Springer International Publishing AG 2017
C. Grueau et al. (Eds.): GISTAM 2016, CCIS 741, pp. 1–14, 2017.
DOI: 10.1007/978-3-319-62618-5_1

landslide inventory mapping studies typically focus on outlining boundaries and neglect the wealth of information revealed by internal deformation features [13].

With regard to deformation analysis, it is possible to consider two main categories [52]: geometrical analysis, which detects the location and the magnitude of the deformation, and physical interpretation, which determines the relationship between the deformation and its causes. In this context, there are four types of models that allow the analysis of deformation [1, 7, 59]. These are the static, kinematic, dynamic and congruence models.

This paper focuses attention on the second type of model. Kinematic models describe deformation as a function of time, including velocity and acceleration. It is also possible to classify data processing techniques into two main groups. The first consists of robust methods, such as Iterative Weighted Similarity Transformation (IWST) and tests (e.g. Chow test, [4, 10]), while the second one is composed of non-robust methods, e.g. Kalman Filter [2, 31, 36, 37, 40, 43, 60]. In this study attention is focused on these last methods, and especially on the Kalman Filter, in order to perform a 3-D deformation analysis using a low-cost single-frequency GPS data in real-time.

Some previous studies have also investigated the accuracy obtainable with geodetic receivers and antennas [38, 39] while some of them have also considered the mass-market ones for landslide monitoring [35, 33, 15]. Those studies used both post-processing [11] and real-time [4] approaches in order to analyze the various types of landslide phenomena. In both cases, the most notable feature of these instruments is that they provide centimeter or sub-centimeter accuracy in real time when the phase ambiguity is fixed as integer value [42]. This is also observed in considering different GNSS positioning techniques [46] such as static [7], rapid-static [32], and real-time kinematic (RTK – [56]) positioning.

A practical case study will show the reliability of the results obtained through the Kalman filter as a statistical tool to detect and predict displacements in real-time. A brief comparison of the results obtained by this method and those obtained with the modified Chow test [4] will conclude this work.

2   GNSS Instruments Considered for Landslides Analysis

Nowadays, many types of GNSS instruments are available functioning in a variety of frequencies, constellations and accuracies obtainable both in real-time and post-processing [17, 16]. As previously stated, GPS/GNSS instruments are very often used for landslides monitoring [18] and are frequently coupled with other instruments such as theodolite, Electronic Distance Measurement (EDM) [30], levels, total station [48], in-clinometers [8], and wire extensometers [6, 12, 28, 41, 45, 53].

In other studies GPS instruments were integrated with other surveying techniques, such as terrestrial laser scanning, Synthetic Aperture Radar (SAR) interferometry [47, 50], and photogrammetry [44], to investigate landslide phenomena [57]. Some studies have also investigated the accuracy of low-cost single-frequency GPS receivers for landslide monitoring [35] both in post-processing [11], and in real-time approaches [4] in order to analyze various types of landslide phenomena (landslides with low constant velocity or with an unexpected and sudden displacement). Considering the first ap-proach, raw GNSS measurements are acquired and post-processed in a single- or

Table 1. Characteristics of GPS receiver and antenna.

Receiver: u-blox LEA EVK-7P Evaluation kit	Antenna: Garmin GA38 GPS & GLONASS L1
Constellation: GPS/GNSS (56 channels)	Constellation: GPS + GLONASS
Observations: C/A L1, Doppler, S/N	Gain 27 dB on the average
Cost: about 200 €	Cost: about 50 €

multi-base solution with one or more GNSS permanent stations or Virtual RINEX while, considering the second one, differential corrections provided by CORSs networks are used to determine the rover position in real-time. In this paper we focus our attention only on this last approach, analyzing displacements in real-time.

The employment of mass-market receivers and antennas is due to the fact that there is an high probability to lose these instruments if an unexpected displacement occurs: in this context, the amount of cost is less than 1000 € for each receiver + antenna that is an order of magnitude less than geodetic instruments. For this study, an u-blox EVK-7P receiver (Table 1 - left) with an external mass-market antenna (Garmin GA38, - Table 1 right) was used. This receiver has a cost of about 350 € (including a patch antenna) while the cost of the Garmin antenna is about 50 €. One of the features of this receiver is that it is possible to use it into a Continuous Operating Reference Stations (CORSs) network.

3 The Tests Performed

The experiments using this mass-market receivers for Network Real Time Kinematic (NRTK) positioning were performed within the SPIN - GNSS positioning service of Regione Piemonte and Regione Lombardia (http://www.spingnss.it/spiderweb/frm Index.aspx) CORSs network (Fig. 1). The network product used is the VRS® stream broadcasted by the network software SpiderNet of the Leica Geosystems® company.

In previous studies [4] it was possible to analyze the performances obtainable considering a static positioning using both real-time and post-processing approaches with this type of receiver and antenna.

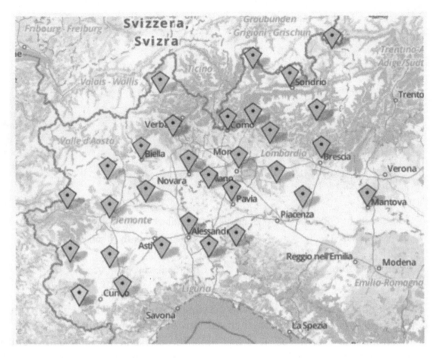

Fig. 1. The SPIN GNSS CORSs network (source: http://www.spingnss.it/spiderweb/frmIndex. aspx).

The antenna was mounted on a sledge (Fig. 2) composed by a complex system of micrometric screws that allow small and controlled displacements, in order to verify the quality of the positioning and the reliability of the statistic tests that were to be performed.

The movements are set by means of a hand-wheel, which moves the sledge along the rail. It is therefore possible with a millimeter tape to obtain direct and visual information about the movements in order to compare the imposed movements against those measured by GNSS instruments. With this sledge, horizontal and vertical movements up to 1.30 m and 0.4 m respectively are possible.

As stated in previous studies [11, 4], there is always a precision of the sledge movement of about 1 mm. Therefore it is possible to consider this value as the "scale resolution" of this support.

The patch antenna was mounted on this sledge as shown in Fig. 2. The positioning results were obtained with a frequency of 1 Hz, considering displacements equal to 1 cm both in planimetry and in altimetry which were provided manually at 30 s intervals.

To perform the NRTK positioning, the routines RTKLIB V. 2.4.2 (http://www. rtklib.com/) were used [54]. This software was chosen because it supports standard and precise positioning algorithms with GPS, GLONASS and Quasi-Zenith.

Fig. 2. The sledge where the GNSS antenna was mounted.

Satellite System (QZSS) constellations, considering also different positioning modalities for both real-time and post-processing approaches: single-point, DGPS/DGNSS, Kinematic, Static, Moving-baseline, Fixed, etc. Moreover the software is able to manage both the proprietary messages (e.g. u-blox LEA-4T, 5T, 6T) and external communication via serial, TCP/IP, NTRIP etc. of several GNSS receivers. In particular, the RTKNAVI tool was used for these experiments. This tool allows the input of both the raw data (pseudorange and carrier-phase measurements) of the u-blox receiver and the stream data coming from a network with NTRIP authentication [58]. For this reason the receiver was connected to a laptop to enable Internet connection and to store the NMEA sentences [42]. Another peculiarity of this software is that it allows the fixing of the phase ambiguity [55] for real-time kinematic positioning, even if the receiver uses only the L1 GPS frequency.

4 The Use of Robust Cluster Analysis to Detect Displacements

Let's assume that a static positioning with a GNSS instrument is performed, considering single-epoch solutions: it means that every second a solution composed by a vector of three coordinates (e.g. latitude, longitude and ellipsoidal height) is obtained. This solution cannot be considered independent from the previous one, as well described in literature [34]. If there are no movements of the receiver it is possible to expect to obtain the same results for every epochs at less than the measurements' noise. So all results can be defined as part of the same cluster.

Starting from this, it is possible to define a variation of the positioning result as an outlier and the same can be made for a displacement: if a displacement occurs, this can be considered an outlier or a sample that it is not a member of the same cluster. This work aims to analyze if it is possible to apply a robust cluster analysis method to detect discontinuities, comparing the obtained results with those available with other statistical robust techniques.

Among the cluster procedures, two different approaches can be followed: the mixture modelling [19] and the trimming approach. This last approach assumes a known fraction α of outliers to be trimmed off and, later, the non-trimmed observations or "regular" data are split into k groups.

One of the main problems of this method is its lack of robustness to outliers detection. Several authors have tried to improve this method thanks to robust estimates. Among these procedures, the "impartial trimming" class is one of the methods most used: these methods are based on the minimization (based on the L_2 norm) of the so called "penality function", that depends by the distance of these k clusters [29]. [14] have developed a method that considers a "trimming function" that minimizes the probability that the rejected data belongs to one of the clusters considered a-priori. [Moreover some recent proposals by [3, 22–27] are made, considering different trimming methods

The methodology used in this work is based on those described in [21] about "spurious outliers model" as a probabilistic framework for robust crisp clustering. Let

$$f(\cdot; \mu, \Sigma) \tag{1}$$

note the probability density function of the *p-variate* normal distribution with mean μ and covariance matrix Σ. The "spurious-outlier model" is defined through "likelihoods" like

$$\left[\prod_{j=1}^{k}\prod_{i\in R_j} f(x_i; \mu_j, \Sigma_j)\right]\left[\prod_{i\in R_0} g_i(x_i)\right] \tag{2}$$

with $\{R_0,\ldots,R_k\}$ being a partition of the set of indices $\{1, 2, \ldots, n\}$ such that $\# R_0 = \lceil n\alpha \rceil$. R_0 are the indices of the "non-regular" observations generated by other (not necessarily normal) probability density functions g_i. "Non-regular" observations can be clearly considered as "outliers" if we assume certain sensible assumptions for the g_i [20, 23, 24]. So the search of a partition $\{R_0,\ldots, R_k\}$ with $\# R_0 = \lceil n\alpha \rceil$, vectors μ_j and positive finite matrices Σ_j maximizing (2) can be simplified as

$$\sum_{j=1}^{k}\sum_{i\in R_j} \log f(x_i; \mu_j, \Sigma_j) \tag{3}$$

Maximizing (3) with k = 1 yields the Minimum Covariance Determinant (MCD) estimator [49]. Unfortunately, the direct maximization of (3) is not a well-defined problem when k > 1. As described in [22], it is easy to see that (3) is unbounded without any constraint on the cluster scatter matrices Σ_j. The *tclust* function, well described in [21], maximizes (3) under different cluster scatter matrix constraints:

(1) on the eigenvalues: based on the eigenvalues of the cluster scatter matrices, a scatter similarity constraint may be defined. With $\lambda_l(\Sigma_j)$ as the eigenvalues of the cluster scatter matrices Σ_j and M_n m_n as the maximum and minimum eigenvalue respectively, it is possible to constrain the ratio M_n/m_n to be smaller or equal than a fixed value ≥ 1. If this value is $= 1$, the *tclust* function tries to solve the trimmed k-means problem as introduced by [14]. This problem simplifies to the well-known k-means clustering criterion when no trimming is done (i.e. $\alpha = 0$). The *tkmeans* function directly implements this most constrained application of the *tclust* function.

(2) on the determinants: this type of constraint limits the relative volumes of the mentioned equidensity ellipsoids, but not the cluster shapes. The use of this type of constraint is particularly advisable when affine equivariance is required.

In this work the *tclust* approach with constraints on eigenvalues and k = 2 are considered: for further details about the formulation of this method, please refer to [21].

5 Data Processing and Results

The method previously described is applied to two real-time datasets that simulates a landslide. As previously stated, the system was composed as shown in Fig. 2. The positioning results were obtained with a frequency of 1 Hz, considering displacements equal to 1 cm both in planimetry and in altimetry provided manually every 30 s. This can be seen in Fig. 3, where two cases, one with good (number of satellites greater than 7) and another one with low (only 5 satellites - Fig. 4) satellite visibility, are considered.

The goal is to estimate if and when a displacement occurs: a starting sample of 10 observations is considered, under the hypothesis that no displacements have been happened during this period. After that, another observation is added to the sample in order to verify if this last observation can be classified as displacement (an outlier of the previous sample). All these analyses are made in real-time considering the results, in terms of coordinates, of NRTK positioning: so this procedure is not applied during the GNSS processing.

When an outlier (or displacement) is discovered, all data of the first cluster are removed and only data belonging to the second cluster are considered, introducing new observations epoch by epoch.

Considering the first kind of data, regarding ah high number of visible satellites, it is possible to affirm that the FS technique is able to detect all displacements imposed thanks to the sledge, as it is possible to see from Fig. 5. This figure shows two different plots: in the upper part there are the results obtained by the NRTK positioning (only for the up component) while in the lower part there are the results of the FS compared to the truth (0 if no displacement has been occurred and 1 if a displacement is happened).

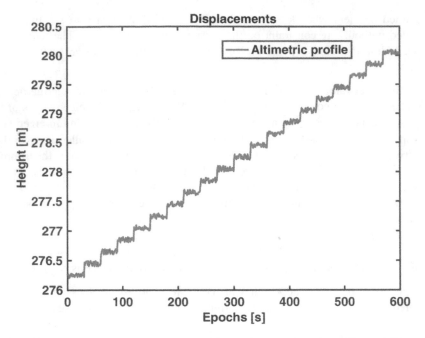

Fig. 3. Altimetric profile of displacements with good (more than 7 satellites - left) satellite visibility.

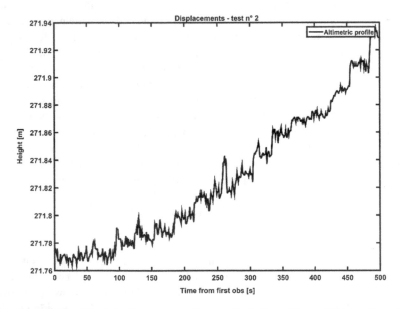

Fig. 4. Altimetric profile of displacements with poor (only 5 satellites - right) satellite visibility considering only epochs with fixed phase ambiguities.

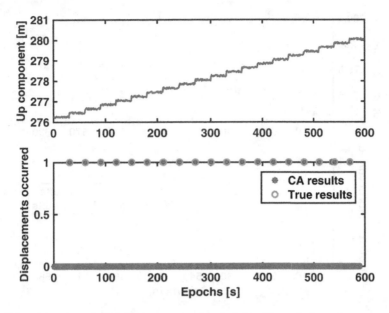

Fig. 5. FS results: in red the displacements detected by this technique, in green the true displacements imposed thanks to the sledge. (Color figure online)

In this case, considering a global trimming level equal to 0.05 ($\alpha = 0.05$) it is detected one more displacement that is not provided manually (epoch 536 - Fig. 6) but this can not be considered as problem: the probability of false alarms (i.e. a reported displacement that has not occurred) is 0.17% that can be considered acceptable.

In this case, this technique has provided the same results obtained with other statistical methods, such as Chow test [4] or a modified version of the Kalman filter [15]: this last method can be considered slightly better because is able to detect all displacements without false alarms.

Considering the global trimming level $\alpha = 0.05$ (Fig. 7), the detected displacements are only 24% with 16 false alarms percentage that is an high number if the number of the total displacements considered is 30.

Even if a sensitivity analysis about α is performed, no substantial results can be obtained: starting from $\alpha = 0.05$ to $\alpha = 0.5$ (that is the maximum value for this parameter, as described in [26]). The best results obtained with this method is with $\alpha = 0.08$ where 53% of discontinuities are detected correctly with an high number (equal to 30) of false alarms.

In this last case, this technique provides worst results if compared to other statistical methods, such as Chow test [4] or a modified version of the Kalman filter [15].

Considering the same dataset, applying Chow test described in the literature [4] and considering a 10-element sample size (meaning that a sample composed of 10 epochs = 10 s of latency of alarm at 1 Hz of acquisition rate), this method correctly identified 93.1% of the displacements, with a very low rate of false alarms equal to 3.3%.

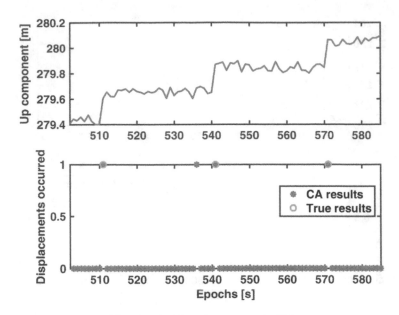

Fig. 6. The false alarm at epoch 536.

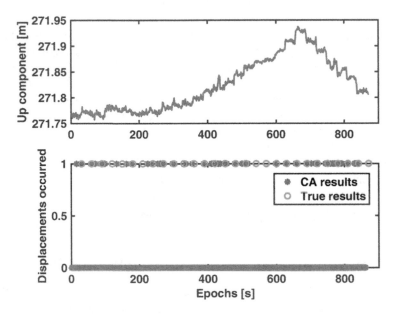

Fig. 7. Results considering real data and $\alpha = 0.05$.

Anyway these values are greater than those estimated with the Kalman filter [15], that has shown the best performances in terms of both false alarms (2.6%) and undetected displacements (1.3%).

6 Conclusions

In this study, a particular robust cluster technique has been applied for kinematic deformation analysis procedure. A GPS data set that simulates a landslide movement was collected and the proposed method has been also tested with real data. The results provided by this technique are finally compared with two other methods (Kalman filter and a modified version of the Chow test). These results are comparable if a simulated dataset is considered, even though the proposed version of the Kalman filter has shown the best performance. If real data are used, the cluster technique has provided worst results if compared to the other two statistical techniques. In the future, other researches about cluster analysis will be made, considering that data are correlated with respect to time and not independent as in the hypothesis of the *tclust* approach. Despite that, interesting results have been obtained in real-time also by employing a mass-market GPS receiver, showing the possibility of using this type of receivers for displacements and deformation analyses.

The employment of these receivers on a landslide site could be useful also from an economic point of view. The total cost of receiver, antenna, transmission system and power supply (solar panel and battery) is about € 600. The advantage is that the economic damage in case of an unexpected event is less than could occur if a geodetic GNSS instrumentation is utilized. At the same time, it is possible to calculate the position of the receivers in a similar way to the CORS network, with obvious advantages in the precision and accuracy of the results and the landslide analysis.

This study has considered fast displacements but it is also possible to suppose a slow motion of the receiver: this aspect will be investigated subsequently.

References

1. Aharizad, N., Setan, H., Lim, M.: Optimized Kalman filter versus rigorous method in deformation analysis. J. Appl. Geodesy 6(3–4), 135–142 (2012)
2. Acar, M., Özlüdemir, M.T., Çelik, R.N., Erol, S., Ayan, T.: Landslide monitoring through Kalman Filtering: a case study in Gürpinar. In: Proceeding of XXth ISPRS Congress, Istanbul, Turkey (2004)
3. Álvarez-Esteban, P.C., Del Barrio, E., Cuesta-Albertos, J.A., Matran, C.: Trimmed comparison of distributions. J. Am. Stat. Assoc. 103(482), 697–704 (2012)
4. Bellone, T., Dabove, P., Manzino, A.M., Taglioretti, C.: Real-time monitoring for fast deformations using GNSS low-cost receivers. Geomatics, Nat. Hazards Risk 7(2), 458–470 (2016)
5. Benoit, L., Briole, P., Martin, O., Thom, C.: Real-time deformation monitoring by a wireless network of low-cost GPS. J. Appl. Geodesy 8(4), 1–10 (2014)
6. Bertachini, E., Capitani, A., Capra, A., Castagnetti, C., Corsini, A., Dubbini, M., Ronchetti, F.: Integrated surveying system for landslide monitoring, Valoria landslide (Appennines of Modena, Italy). Paper presented at: FIG working week 2009, Eilat, Israel (2009)
7. Brückl, E., Brunner, F.K., Lang, E., Mertl, S., Müller, M., Stary, U.: The Gradenbach Observatory - monitoring deep-seated gravitational slope deformation by geodetic, hydrological, and seismological methods. Landslides 10, 815–829 (2013)

8. Calcaterra, S., Cesi, C., Di Maio, C., Gambino, P., Merli, K., Vallario, M., Vassallo, R.: Surface displacements of two landslides evaluated by GPS and inclinometer systems: a case study in Southern Apennines. Italy. Nat. Hazards **61**(1), 257–266 (2012)

9. Chrzanowski, A., Chen, Y., Romero, P., Secord, J.M.: Integration of geodetic and geotechnical deformation surveys in the geosciences. Tectonophysics **130**(1–4), 369–383 (1986)

10. Chow, G.C.: Tests of equality between sets of coefficients in two linear regressions. Econometrica **28**(3), 591–605 (1960)

11. Cina, A., Piras, M.: Monitoring of landslides with mass market GPS: an alternative low cost solution. Geomatics, Nat. Hazards Risk (2014). http://www.tandfonline.com/doi/full/10. 1080/19475705.2014.889046#.U0-vV6IzfcB. Accessed Feb 2014

12. Coe, J.A., Ellis, W.L., Godt, J.W., Savage, W.Z., Savage, J.E., Michael, J.A., Kibler, J.D., Powers, P.S., Lidke, D.J., Debray, S.: Seasonal movement of the Slumgullion landslide determined from Global Positioning System surveys and field instrumentation, July 1998-March 2002. Eng. Geol. **68**(1–2), 67–101 (2003)

13. Cruden, D.M.: A simple definition of a landslide. Bull. Int. Assoc. Eng. Geology (Bulletin de l'Association Internationale de Géologie de l'Ingénieur) **43**(1), 27–29 (1991)

14. Cuesta-Albertos, J., Gordaliza, A., Matràn, C.: Trimmed k-means: an attempt to robustify quantizers. Ann. Stat. **25**(2), 553–576 (1997)

15. Dabove, P., Manzino, A.M.: Kalman Filter as tool for the real-time detection of fast displacements by the use of low-cost GPS receivers. In: Proceedings of the 2nd International Conference on Geographical Information Systems Theory, Applications and Management, pp. 15-23 (2016). ISBN 978-989-758-188-5

16. Dabove, P., Manzino, A.M., Taglioretti, C.: GNSS network products for post-processing positioning: limitations and peculiarities. Appl. Geomatics **6**(1), 27–36 (2014). http://link. springer.com/article/10.1007/s12518-014-0122-3, ISSN 1866-9298

17. Dabove, P., Manzino, A.M.: GPS & GLONASS mass-market receivers: positioning performances and peculiarities. Sensors **14**, 22159–22179 (2014)

18. Eyo, E.E., Musa, T.A., Idris, K.M., Opaluwa, Y.D.: Reverse RTK data streaming for low-cost landslide monitoring. In: Abdul Rahman, A., Boguslawski, P., Anton, F., Said, M.N., Omar, K.M. (eds.) Geoinformation for Informed Decisions. LNGC, pp. 19–33. Springer, Cham (2014). doi:10.1007/978-3-319-03644-1_2

19. Fraley, C., Raftery, A.E.: How many clusters? Which clustering method? Answers via model-based cluster analysis. Comput. J. **41**(8), 578–588 (1998)

20. Fritz, H., García-Escudero, L.A., Mayo-Iscar, A.: tclust: An R package for a trimming approach to cluster analysis. J. Stat. Softw. **47**(12), 1–26 (2012)

21. Fritz, H., García-Escudero, L.A., Mayo-Iscar, A.: A fast algorithm for robust constrained clustering. Comput. Stat. Data Anal. **61**, 124–136 (2013)

22. Gallegos, M.T.: Robust clustering under general normal assumptions (2001). http://www. fmi.uni-passau.de/forschung/mip-berichte/MIP-0103.html

23. Gallegos, M.T.: Maximum likelihood clustering with outliers. In: Jajuga, K., Sokołowski, A., Bock, H.H. (eds.) Classification, Clustering, and Data Analysis. Studies in Classification, Data Analysis, and Knowledge Organization, pp. 247–255. Springer, Heidelberg (2002). doi:10.1007/978-3-642-56181-8_27

24. Gallegos, M.T., Ritter, G.: A robust method for cluster analysis. Ann. Stat. **33**, 347–380 (2005)

25. Garcia-Escudero, L.A., García-Escudero, A., Matrán, C., Mayo-Iscar, A.: A general trimming approach to robust cluster analysis. Ann. Stat. **36**, 1324–1345 (2008)

26. García-Escudero, L.A., Gordaliza, A., Matrán, C., Mayo-Iscar, A.: A review of robust clustering methods. Adv. Data Anal. Classif. **4**(2–3), 89–109 (2010)

27. García-Escudero, L.A., Gordaliza, A., Matrán, C., Mayo-Iscar, A.: Exploring the number of groups in robust model-based clustering. Stat. Comput. **21**(4), 585–599 (2011)
28. Gili, J.A., Corominas, J., Rius, J.: Using global positioning system techniques in landslide monitoring. Eng. Geol. **55**(3), 167–192 (2000)
29. Gordaliza, A.: Best approximations to random variables based on trimming procedures. J. Approx. Theory **64**, 162–180 (1991)
30. Günther, J., Heunecke, O., Pink, S., Schuhbäck, S.: Developments towards a low cost GNSS Based Sensor Network for the monitoring of landslides. Paper presented at: 13th FIG International Symposium on Deformation Measurements and Analysis, Lisbon (2008)
31. Gülal, E.: Application of Kalman filtering technique in the analysis of deformation measurements. J. Yıldız Technical University, (1) (1999). (in Turkish)
32. Hastaoglu, K.O., Sanli, D.U.: Monitoring Koyulhisar landslide using rapid static GPS: a strategy to remove biases from vertical velocities. Nat. Hazards **58**, 1275–1294 (2011)
33. Heunecke, O., Glabsch, J., Schuhbäck, S.: Landslide monitoring using low cost GNSS equipment – experiences from two alpine testing sites. J. Civil Eng. Archit. **45**, 661–669 (2011)
34. Hoffman-Wellenhof, B., Lichtenegger, H., Wasle, E.: GNSS-Global Navigation Satellite Systems. GPS, GLONASS Galileo and More. Springer, Wien (2008)
35. Janssen, V., Rizos, C.: A mixed-mode GPS network processing approach for deformation monitoring applications. Surv. Rev. **37**(287), 2–19 (2003)
36. Kalman, R.E.: A new approach to linear filtering and prediction problems. Trans. ASME J. Basic Eng. **82**, 35–45 (1960)
37. Li, L., Kuhlmann, H.: Detection of deformations and outliers in real-time GPS measurments by Kalman Filter Model with Shaping Filter. In: 4th IAG Symposium on Geodesy for Geotechnical and Structural Engineering and 13th FIG Symposium on Deformation Measurements, Lisbon (2008)
38. Li, L., Kuhlmann, H.: Deformation detection in the GPS real-time series by the Multiple Kalman Filter Model. J. Surveying Eng. **136**, 157–164 (2010)
39. Li, L., Kuhlmann, H.: Real-time deformation measurements using time series of GPS coordinates processed by Kalman Filter with Shaping Filter. Surv. Rev. **44**(326), 189–197 (2012)
40. Mäkilä, P.M.: Kalman Filtering and Linear Quadratic Gaussian Control. Lecture notes for course 7604120, Part I (2004). http://www.dt.fee.unicamp.br/~jbosco/ia856/KF_part1_Makila.pdf
41. Malet, J.-P., Maquaire, O., Calais, E.: The use of global positioning system techniques for the continuous monitoring of landslides: application to the super-sauze earth flow (Alpes-de-Haute-Province, France). Geomorphology **43**, 33–54 (2002)
42. Manzino, A.M., Dabove, P.: Quality control of the NRTK positioning with mass-market receivers. In: Hsueh, Y.-H. (ed.) Global Positioning Systems: Signal Structure, Applications and Sources of Error and Biases (Chap. 2), pp. 17–40. Hauppauge NY, New York (2013)
43. Masiero, A., Guarnieri, A., Vettore, A., Pirotti, F.: A nonlinear filtering approach for smartphone-based indoor navigation, Tainan, Taiwan, 1–3 May 2013
44. Mora, P., Baldi, P., Casula, G., Fabris, M., Ghirotti, M., Mazzini, E., Pesci, A.: Global positioning systems and digital photogrammetry for the monitoring of mass movements: application to the Ca' di Malta landslide (Northern Apennines, Italy). Eng. Geol. **68**(1–2), 103–121 (2003)
45. Moss, J.L.: Using the global positioning system to monitor dynamic ground deformation networks on potentially active landslides. Int. J. Appl. Earth Obs. Geoinf. **2**(1), 24–32 (2000)

46. Othman, Z., Wan Aziz, W.A., Anuar, A.: Evaluating the performance of GPS survey methods for landslide monitoring at hillside residential area: static vs rapid static. In: IEEE 7th International Colloquium on Signal Processing and its Applications, George Town, Penang (2011)
47. Peyret, M., Djamour, Y., Rizza, M., Ritz, J.F., Hurtrez, J.E., Goudarzi, M.A., Nankali, H., Chery, J., Le Dortz, K., Uri, F.: Monitoring of the large slow Kahrod landslide in Alboz mountain range (Iran) by GPS and SAR interferometry. Eng. Geol. **100**, 131–141 (2008)
48. Rizzo, V.: GPS monitoring and new data on slope movements in the Maratea Valley (Potenza, Basilicata). Phys. Chem. Earth, Parts A/B/C **27**(36), 1535–1544 (2002)
49. Rousseeuw, P.J.: Multivariate estimation with high breakdown point. In: Vincze, I., Grossmann, W., Pflug, G., Wertz, W. (eds.) Mathematical Statistics and Applications, vol. B, pp. 283–297. Reidel, Dordrecht (1985)
50. Rott, H., Nagler, T.: The contribution of radar interferometry to the assessment of landslide hazards. Adv. Space Res. **37**(4), 710–719 (2006)
51. Sedlak, V., Jecny, M.: Deformation measurements on Bulk Dam of waterwork in East Slovakia. Geol. Ecol. Min. Serv. **L**(2), 1–10 (2004)
52. Szostak-Chrzanowski, A., Chrzanowski, A., Massiéra, M.: Use of deformation monitoring results in solving geomechanical problems—case studies. Eng. Geol. **79**(1), 3–12 (2005)
53. Tagliavini, F., Mantovani, M., Marcato, G., Pasuto, A., Silvano, S.: Validation of landslide hazard assessment by means of GPS monitoring technique – a case study in the Dolomites (Eastern Alps, Italy). Nat. Hazards Earth Syst. Sci. **7**, 185–193 (2007)
54. Takasu, T., Yasuda, A.: Development of the low-cost RTK GPS receiver with the open source program package RTKLIB. In: International Symposium on GPS/GNSS, International Convention Centre, Jeju, Korea (2009)
55. Teunissen, P.J.G., Salzmann, M.A.: A recursive slip-page test for use in state-space filtering. Manuscripta Geodaetica **14**, 383–390 (1989)
56. Wang, G.: GPS landslide monitoring single base vs. network solutions—a case study based on the Puerto Rico and Virgin Islands permanent GPS network. J. Geodetic Sci. **1**(3), 191–203 (2011)
57. Wang, G., Soler, T.: OPUS for horizontal sub-centimeter accuracy landslide monitoring: case study in Puerto Rico and Virgin Islands region. J. Surv. Eng. **138**(3), 11 (2012)
58. Weber, G., Dettmering, D., Gebhard, H.: Networked transport of RTCM via internet protocol (NTRIP). In: International Association of Geodesy Symposia: A Window on the Future of Geodesy, vol. 128 (2006)
59. Welsch, W., Heunecke, O.: Models and terminology for the analysis of geodetic monitoring observations. Official report of the ad-hoc committee of FIG working group, vol. 6, pp. 390–412 (2001)
60. Wei, Z., Dongli, F., Jinzhong, Y.: Adaptive Kalman Filtering method to the data processing of GPS deformation monitoring. In: Proceedings of the 2010 International Forum on Information Technology and Applications. - Volume 01 (IFITA 2010), vol. 1, pp. 288–292. IEEE Computer Society, Washington, DC, USA (2010). doi:10.1109/IFITA.2010.18

Integrating a Metadata Editor with Hyperbolic Tree to Improve Data Access in Spatial Data Infrastructures

Marcos Vinícius Montanari, Alexandra Moreira, Vitor Eduardo Concesso Dias,
Eduardo Lourenço, and Jugurta Lisboa-Filho[✉]

Department of Informatics, Federal University of Viçosa, Viçosa, MG, Brazil
{marcos.montanari,vitor.dias,eduardo.lourenco,jugurta}@ufv.br,
xandramoreira@yahoo.com.br

Abstract. The storage and management of metadata is an essential task in systems that deal with spatial data. However, to accomplish this task it is necessary to count with a set of tools that operate in integrated mode. This paper describes the development of edpMGB, a metadata editor for the MGB profile. The editor is open-source and is being released in the cloud via Web following the Software as a Service (SaaS) model, so it can be accessed from any site on the Internet, requiring only a Web browser. The editor was developed on the scope of the research and development project (R&D) Geoportal Cemig SDI-based corporate GIS (Geoportal Cemig - SIG corporativo baseado em IDE), whose objective is the implementation of a corporate SDI for the Minas Gerais Power Company (Companhia Energética de Minas Gerais - Cemig). Furthermore, this work proposes the use of a hyperbolic tree to aid indexing metadata, facilitating its recovery. After indexing, the user can navigate through the tree nodes and perform searches over the metadata related to the search words.

Keywords: Spatial Data Infrastructure · INDE · Editor · Metadata · MGB profile

1 Introduction

Within Information and Communication Technology, information redundancy and lack of data standardization proves very common. The same data often ends up being produced, managed, used, and stored by several independent producers that use different formats and standards, who seek to meet exclusively the individual needs of specific users [8].

The greater production of spatial data requires documenting them so they can be reused. A piece of data immersed in its context becomes information, however, with no such documentation, it is virtually worthless information [22].

In order to prevent actions in duplicity and wasted resources to obtain spatial data, the Brazilian government began, in 2003, studies aiming to integrate and

© Springer International Publishing AG 2017
C. Grueau et al. (Eds.): GISTAM 2016, CCIS 741, pp. 15–31, 2017.
DOI: 10.1007/978-3-319-62618-5_2

reuse geo-spatial data produced by the different federal administration organs. In 2008, Act 6,666 of November 27th established the National Spatial Data Infrastructure (Infraestrutura Nacional de Dados Espaciais - INDE) [3]. INDEs goal is "to catalog, integrate, and harmonize the geospatial data produced and maintained by the different governmental institutions so as to facilitate their location, exploration, and access by any user connected to the Internet" [7].

To [18], geographic metadata correspond to the documentation of geographic data and are created according to standards. Such standards consist of a set of regulations that allow the geographic data to be described textually in a previously established manner. The most well-known geographic metadata standards were defined by the Federal Geographic Data Committee (FGDC) and by the International Organization of Standards (ISO). These institutions established international geographic metadata standards that meet the needs of different users, thus which comprises the variability in geographic information [25].

In order to meet the metadata standardization demands started by INDE, the National Cartography Committee (CONCAR) created the Brazilian Geospatial Metadata Profile (MGB Profile) based on norm ISO 19115:2003. A metadata profile is a basic set of elements that portray the characteristics of geospatial products of a given community and guarantees their identification [15].

Based on norm ISO 19139:2007, XML schemas were defined to materialize and code ISO 19115:2003 as a file. Since the MGB profile is based on norm ISO 19115:2003, its metadata must also follow the materialization standards defined in norm 19139:2007 to increase the interoperability among the systems that use the profile as their base [24].

The present paper aims to describe the development of edpMGB, a metadata editor for the MGB profile. edpMGB is open-source software and is available on the cloud via the web following the model Software as a Service (SaaS). Thus, it can be accessed from anywhere and requires only a web browser [30]. Some features of this metadata editor are also available through web services, which allow other systems to use some of this tools features. The editor was developed in the context of the research and development (R&D) project "GeoPortal Cemig SIG corporativo baseado em IDE" (Cemig GeoPortal SDI-based corporate GIS) being developed to help implement a corporate SDI for Minas Gerais Power Company - Cemig (Compania Energética de Minas Gerais).

The main reason for developing edpMGB was the need for a specific tool for this metadata standard focusing on the Brazilian technical audience that works with spatial information. Other editors are available in the market, such as CatMDEdit [6] and Geonetwork [13]. Although these are metadata editors, they are not specific for documenting metadata in the MGB profile. Geonetwork, in particular, can be used alongside the editor described in the present study.

The metadata registered by the editor certainly have a position in the conceptual network underlying the application domain. In order to establish this position, it can use some mechanism to browse the network domain related words. Here we propose the use of Hyperbolic Tree [17] on a network domain words to assist in indexing the registered metadata.

2   Theoretical Framework

This section describes the theories, techniques and tools on which this research sought support, starting on the description of the concepts of Spatial Data Infrastructure and metadata for later addressing standards and related techniques.

2.1   Spatial Data Infrastructure

According to [22], the term "Spatial Data Infrastructure" (SDI) refers to the collection of relevant technologies, policies, and institutional arrangements that facilitate providing and accessing spatial data. It provides a base for discovering geographic data, besides the evaluation and application for users and service providers at all governmental levels, of the commercial sector, of non-profit organizations, universities, and citizens as a whole. Moreover, it hosts the geographic data, attributes, documentation (metadata), and some methods to access such data. A functional SDI must include the organizational agreements required to coordinate and administer it at a local, regional, national, and/or global level in order to enable the ideal environment to interconnect applications to data.

According to [26], the main components of an SDI are:

- Institutional platform related with the policies and administrative agreements in the implementation of standards and data;
- Technical standards that define the technical characteristics of the main data;
- Access network, which make the data accessible by users;
- Data produced in the institutional platform and that meet the technical standards;
- People, comprising users, data producers, and any agent that adds value to the SDI development process.

Figure 1 shows the relation among the elements of an SDI considering people and data as one category and access networks, policy, and technical norms as another. The relation among the categories is very dynamic given the changes in the communities and their needs, as well as the requirement of different datasets.

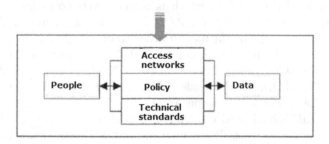

Fig. 1. Nature and relations among the components of an SDI [24].

The importance of the SDI is translated as the potential for sharing geographic information that integrates several types of users. It aims to decrease the lack of interoperability among different systems by standardizing the data. It is a simple concept, but its implementation is hard in view of the technological, political, and cultural differences among the geographic data producers [20].

2.2 Metadata

The increasing production of spatial data furthers the need to document such data so they can be used in future studies. A piece of data with its context (documentation and metadata) becomes information, whereas, without such details, it is virtually worthless information [22].

Metadata represent the information on a piece of data and can be exemplified by descriptions about who, how, when, and why the data was produced. According to Decree-Law no. 6,666 of November 27th, 2008 enacted by the Brazilian Federal Executive Branch, metadata represents the: "Set of descriptive information on the data, including the characteristics of its surveying, production, quality, and storage structure, essential to promote its documentation, integration, and availability, as well as enable searching and browsing them." Metadata use is related to the needs an organization has of better knowing its stored data in more detail. Data cataloging will facilitate its use and meet the users needs. Without this efficient documentation, the users have more difficulty in locating the data required for their applications. This form of data representation aims to contribute to orienting, developing, and describing electronic documents by creating standards, production, and handling in the description by metadata [27].

According to [3], each organ and entity of the Brazilian federal government must share the geospatial metadata they produce. To this end, standards must be used when creating metadata since they are being generated by different organs, however, following the same guidelines to ensure interoperability.

Therefore, the National Cartography Commission (Comissão National de Cartografia CONCAR) defined in 2009 the Geospatial Metadata Profile of Brazil (Perfil de Metadados Geoespaciais do Brasil MGB Profile) in order to establish a national standard used to create geospatial metadata. The profile was developed by several organs that produce geospatial data in Brazil based on international norm ISO 19115:2003. However, this profile does not define how these metadata are structured in electronic files, which is managed by norm ISO 19139:2007, which aims to materialize the concepts of ISO 19115:2003 into a file by coding these metadata in XML schemas. In order to create the MGB profile, some established profiles, also based on norm ISO 19115, were analyzed: MIG Metadados de Informação Geográfica (Portugal); NEM Núcleo Español de Metadados (Spain); NAP North American Profile (EUA/Canadá); LAMP Latin American Metadata Profile (proposed for Latin America) [24].

Norm ISO 19115:2003 used the Unified Modeling Language (UML) to represent the metadata sections. It featured around 400 elements, eight of which mandatory, for profiles derived from this standard. This norm allows defining profiles and extensions for specific application fields [7].

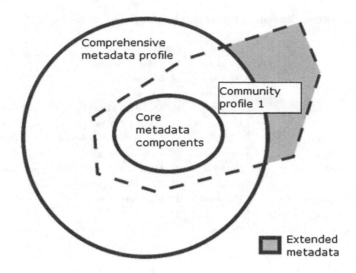

Fig. 2. Metadata profile of a community [6].

A metadata profile is a set of elements that meets the needs of a given community and guarantees their identification. A metadata set defined by the standard may not be enough to meet the needs of a given community, hence the profile may allow additional elements to be used to fulfill the users needs. Figure 2 shows the relation between the metadata set by the standard, the core components, the profile of a community, and the extension made to this profile [7].

The MGB profile has a version called "summarized MGB profile", which represents the minimum set of elements the geospatial metadata produced must have [7]. The abridged profile has 23 elements, presented in Table 1. The user has the option of choosing between the complete version, with 82 elements, and the abridged version. The elements in the MGB profile are organized into the following sections:

- Section 1 (Identification) - provides basic information on the geospatial dataset such as title, date, person or institution responsible, and summary of the data;
- Section 2 (Identification of the geographic dataset GDS) - refers to the information required to identify and evaluate a GDS. This section characterizes the type of spatial representation, scale, language, extension, etc.;
- Section 3 (Restriction information) - publicizes information regarding access and use restrictions and is made up of two entities, one regarding legal restrictions and the other, security restrictions;
- Section 4 (Quality) - allows the quality of a dataset to be evaluated by informing the hierarchical level, linage, and report on the data;
- Section 5 (Maintenance information) - informs the maintenance and update frequency;

Table 1. Entities and elements of the metadata core of the summarized MGB profile [6].

Entity/Element	Condition	Entity/Element	Condition
1. Title	Mandatory	13. Reference system	mandatory
2. Date	Mandatory	14. Linage	optional
3. Responsible	Mandatory	15. Online access	Optional
4. Geographic extension	Conditional	16. Metadata identifier	Optional
5. Language	Mandatory	17. Standard metadata name	Optional
6. Character encoding	Conditional	18. Metadata norm version	Optional
7. Thematic category	Mandatory	19. Metadata language	Conditional
8. Spatial resolution	Optional	20. Metadata character encoding	Conditional
9. Summary	Mandatory	21. Responsible for the metadata	Mandatory
10. Distribution format	Mandatory	22. Metadata date	Mandatory
11. Time and altimetry extension	Optional	23. Status	Mandatory
12. Type of spatial representation	Optional		

- Section 6 (Spatial representation information) - describes the mechanisms used to represent the spatial information (matrix or vector);
- Section 7 (Reference system) - information on the reference system, including the coordinates system and the geodesic referential of the spatial dataset;
- Section 8 (Content information) - describes the catalog of features and the content of the matrix data;
- Section 9 (Distribution) - reports information related to the distributor and to the alternatives to obtain geographic data;
- Section 10 (Metadata) - section responsible for information on its own metadata. It includes the person or institution responsible, creation date, norm used, etc.

Norm ISO 19139:2007 defines a set of XML schemas for metadata defined in ISO 19115:2003. It aims to define a file format for geospatial metadata that follows ISO 19115:2003. These schemas enable structuring and validating the metadata's XML files in accordance with the norm [16]. The XML format allows the metadata instances to go around the Internet, including geospatial web services, as predicted in the specification of the Web Services Catalog [23].

Maintained since 1998 by the W3C, XML is a flexible and simple mark-up language. Its main characteristics are being a text-based language, separating content from formatting, being simple and easily interpreted, and allowing the creation of limitless tags, which facilitates online data exchange [14]. An XML document can be considered well formatted if it matches what it prescribed in the norms [4]. This document can be valid as long as it follows some norms described in its grammar. The XML Schema Definition (XSD) is an XML grammar format [10]. ISO 19139 is described as XML schemas built under the specification by [28].

2.3   Recovering and Visualizing Information

For [29], in communication, an emitter uses mechanisms of information to act aiming to transmit knowledge, which, in turn, must be absorbed by a receiver.

This relation is only possible by using language. Thus, the study of the recovery process involves knowledge in logic, technology, and linguistics.

As information production increased over the years, means were developed to facilitate the process of recovering information to aid the needs of users of both traditional and digital libraries. One such technique to aid recovering information is the use of indexes as a collection of content that indicate where the information the user seeks can be found. Such content can be organized to facilitate search [2]. An example of the importance and use of indexes are in content searches in encyclopedias in which indexes are used to direct the user to the content of their interest without having to read or access other material content.

As the amount of information grows, the complexity of the objects stored also grew and the large volume of data began requiring increasingly enhanced recovery techniques. In face of this new reality, the information recovery process becomes increasingly more important [5].

For [2], the indexing process pertains to the creation of data structures associated with the textual portion of documents such as suffix array and inverted file structures, for example. Those structures may contain data on the characteristics of the terms in the document collection such as the number of times each term appears in a given document.

Information recovery systems present the content found in the corpus and allow the items that meet the users needs to be chosen using a search expression. A simplified representation of the information recovery process is shown in Fig. 3.

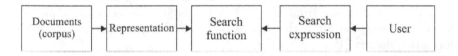

Fig. 3. Representation of the information recovery process [11].

Within the information recovery process is the search function, which compares the search expression provided by the users and the documents that are part of the corpus to return the items that feature the information the user seeks. However, even if a term in the search expression features in the representation of a document, that does not mean this document is useful to the user [11].

In this process of information retrieval, data visualization can help in the search carried out by the user. Visualizing information consists in using computing resources to help analyze and understand a dataset. This help is mostly provided by using visual forms [12].

This type of visualization studies the ways of transforming abstract data into real or mentally visible images that facilitate understanding them and helps obtain new pieces of information contained in the data. This process aids in understanding a subject that would be hard to understand otherwise [21].

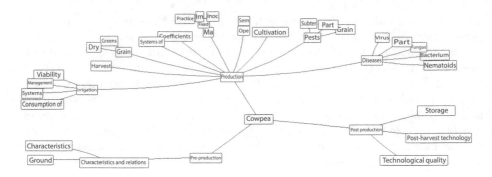

Fig. 4. Example of hyperbolic tree [13].

The visualization techniques represent data graphically to facilitate under-standing them. Such representations can be split into three classes: unidimen-sional, bidimensional, or tridimensional, according to the space dimension where the geometric elements used are located [12].

2.4   Hyperbolic Tree

The hyperbolic tree (Fig. 4) is a graphical visualization resource that facilitates navigation in very large data hierarchies. In this type of representation, the base is the visual form of a tree structure within a hyperbolic space bounded by a circular region. The center of the circle features the observation focus and, the greater the distance from the center, the larger the number of pieces of information presented, while the regions close to the edge of the circle have high information density [12].

This technique is widely used in sitemaps and as a navigation tool in several hierarchies. Links to other pages can be assigned to its nodes. Node color can also be customized [9].

2.5   Cloud Computing and Web Services

According to [1], cloud computing refers to applications provided as online ser-vices as well as the hardware and software systems that provide such services. These systems favor the user's access to different applications using the Inter-net, regardless of the platform used or where the user is. Therefore, all data processing and storage occurs at some online datacenter, as virtual servers.

Cloud computing takes place through different types of services. The main one, which will be approached in this study, is the concept of Software as a Service (SaaS) [31]. This type of service has several resources that the end user can access via web browsers [1]. The metadata editor presented in this study is an example of SaaS.

According to [30], a web service is a software system designed to support the interoperability among machines on a network. A web service is accessed through platform-independent protocols and data formats such as the Hypertext Transfer Protocol (HTTP), eXtensible Markup Language (XML), and Simple Object Access Protocol (SOAP). A web service's interface is accessible through standardized XML messages, i.e., in text format [19].

The editor hereby presented makes available web services that can be accessed via SOAP by any application regardless of the technology used, with no need to access the system's interface. The services can be used by developers that want to use them in their applications.

3   Related Works

Filling out metadata is hard work, as is any other product cataloging.

The use of software to generate geographic metadata files in several formats that meet different norms has been proposed to improve geographic data documentation. Among these tools, the freeware ones such as CatMDEdit and Geonetwork stand out, both featuring forms to fill out the metadata according to a predefined profile and enabling the automatic extraction of some characteristics from the dataset.

3.1   CatMDEdit

CatMDEdit [6] is a tool to edit and visualize metadata in several standards that facilitates resource documentation, particularly geographic information. It is developed by the Instituto Geográfico Nacional de España (IGN) along with the Advanced Information Systems Group (IAAA) from the University of Zaragoza, with technical support from the GeoSpatiumLab (GSL). The tool was implemented in Java and has important features for metadata documentation, among which: The system is multi-platform (running on Linux and Windows); multilingual (Spanish, English, French, German, Polish, Portuguese, and Czech); open source, supporting the automated metadata file extraction and generation (Shapefile, DGN, ECW, FICC, GeoTIFF, GIF/GFW, JPG/JGW, PNG/PGW); and it converts and personalizes metadata standards to generate new metadata standards and profiles in order to serve all types of geographic data.

3.2   GeoNetwork

Geonetwork [14] is a standardized decentralized environment based on a catalog system to facilitate geospatial data access, recovery, update, and management. It provides a complete environment with metadata editor and catalog with search functions. It also carries an online interactive map viewer using web map service. It is currently used in countless SDI initiatives worldwide. Some of its main features include: (1) native support to the metadata standards ISO 19115, ISO 19139, FGDC, and Dublin Core, besides being able do configure a new metadata standard profile; (2) metadata synchronization among distributed catalogs; (3) user management and customized access control; (4) cataloging and access to several types of data and documents (upload/download); (5) interface with multilingual support; and (6) metadata importing in the ISO 19115 standard into a metadata profile configured in Geonetwork. This system is free and open-source, which facilitates its evolution and customization by SDI developers.

What differentiates edpMGB from the editors mentioned in this section is that the former is a SaaS system that does not need to be installed on the users machine, besides having a simplified interface that helps users document metadata. Moreover, it is the only editor specific for the MGB profile and its interface has the local advantage of being in Portuguese.

4   edpMGB - MGB Profile Metadata Editor

Metadata documentation in the MGB profile is an important task for them to be shared and reused. Up until now, no other editor specific for the MGC profile has been created, hence, each metadata set is specified according to the preferences of their authors [24].

edpMGB is a web application developed with the Google Web toolkit (GWT), a framework developed by Google for web-applications. The GWT used the java programming language to develop applications and the tool itself converts the JavaScript code so that the application can be interpreted by any web browser regardless of the platform the user is running.

Through edpMGB, the user can create, edit, and save metadata as a .xml file following international standards so it can be used in several geospatial tools. Before the XML file is generated, it must be validated according to the MGB profile rules. One of the components of GeoPortal Cemig is the geospatial metadata catalog. Therefore, the metadata are documented through edpMGB, thus integrating SDI-Cemig to the INDE.

Figure 5 illustrates edpMGBs home screen. The left-hand side has the navigation tree separated into sections and profile elements. The center-right area features the screen with the MGB profile fields divided into ten panels that represent the sections, which may be accessed using the green arrows to the right or to the left.

Figure 6(a) shows the rules of enforcement, occurrence, and type of value of an element. The icon shown in Fig. 6(b) shows an elements detailed information. The bottom part of the home screen features the editor buttons panel. The button "Abrir" (Open) displays the dialog box where the user can load an XML file to be edited. Each element of the profile will be loaded in its respective text box.

The button "Limpar" (Clear) clears the open elements text box, while "Limpar Tudo" (Clear All) clears all text boxes of the metadata being edited.

The button "Validar" (Validate) performs one of the system's main features, which is to validate whether the metadata is in accordance with the XML schema of the MGB profile. When the button is clicked, the system verifies the data input and may display, for instance, a dialog box as shown in Fig. 7 to inform that the metadata does not respect the MGB profiles rules. The dialog box displays an

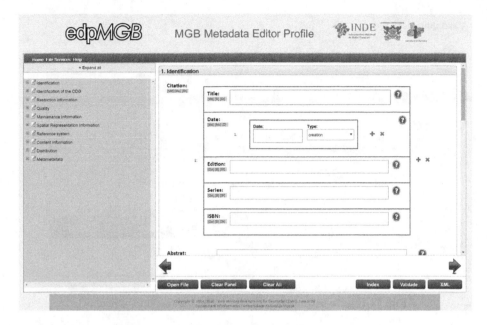

Fig. 5. edpMGBs home screen. (Color figure online)

Fig. 6. Warning dialog about mandatory elements.

error log alerting the user and showing which fields are not in accordance with the rules. The user may keep editing the metadata or store it in his or her machine even if it is not validated for the MGB profile. When a non-validated metadata is generated, it is tagged informing it does not conform to the MGB profile.

A `generateXMLScript()` method receives as parameter the elements of the MGB profile, which are validated by the method `validateMGB()`, then the former outputs the XML script as a string.

The method `validateMGB()` validates the metadata according to the MGB profile, receiving its elements as parameter and outputting a list with the error messages found in the metadata or an empty list in case it respects the profile.

To [24], most metadata provided by national data producers do not fully respect the profile's rules, which is a big issue since it compromises the interoperability among the systems that use the same profile. Nonetheless, the impossibility of saving a non-validated metadata may cause problems to users, perhaps due to the lack of information on the metadata elements. Hence, the user has the

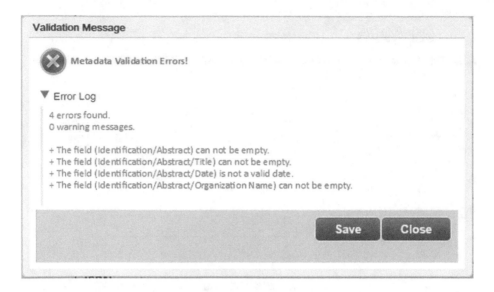

Fig. 7. Result with errors found by the validation service.

option of saving the metadata even if it does not conform to the MGB profile and, if needed, load it again in the tool for further editing. The "XML" button saves the metadata as a .xml file (Fig. 8) that contains all information input in the fields. The .xml file is saved in the users computer.

5   Hyperbolic Tree for the Electric Sector

The edpMGB editor has a function to index the metadata produced and validated by it through a network of words displayed as a hyperbolic tree created for the electricity sector. The Sect. 5.1 explains how the network of words displayed by the hyperbolic tree was built and the way the metadata is indexed is explained in the Subsect. 5.2.

5.1   Extracting Terms Used in the Electric Sector

The terms used to create the hyperbolic tree were extracted using the tool EχATOLP Extrator Automático de Termos para Ontologias em Língua Portuguesa, developed to extract terms from linguistically annotated corpora [19]. EχATOLP is a tool that takes a corpus and automatically extracts all nominal syntagmas (NS) in the text while classifying them according to the number of words. The syntagmas extracted are saved in lists that may have both the NSs in their original form in the text and in its canonical form. The tool also provides some common options for manipulating lists of terms such as applying cut-off points, comparing lists, and calculating usual measures of precision and scope [19].

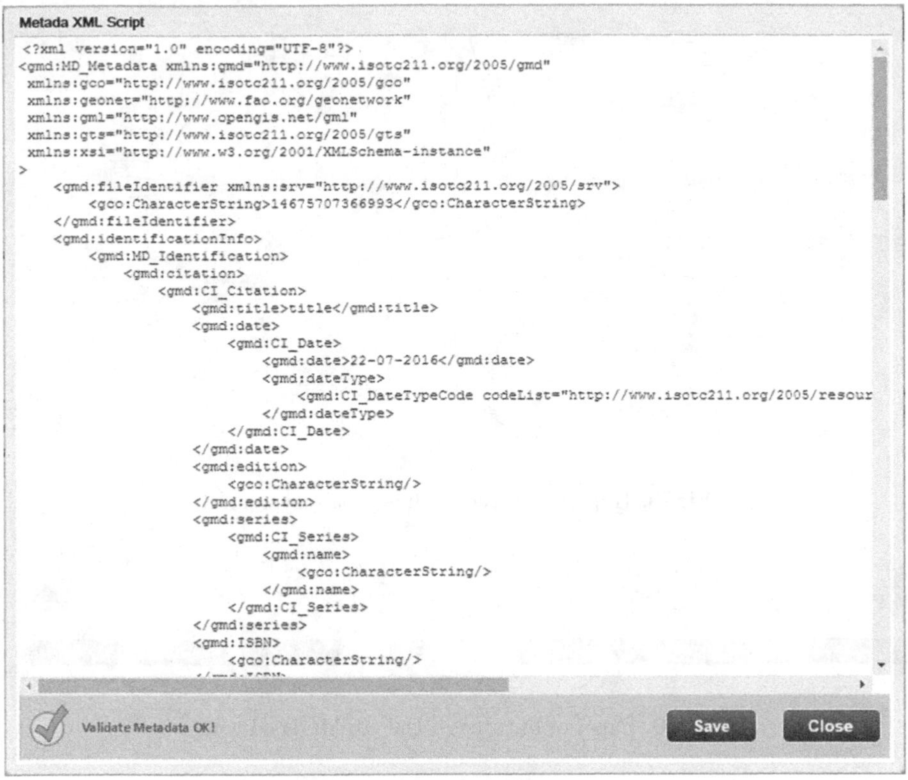

Fig. 8. XML file generation screen.

A domain corpus is a set of texts on a specific field of knowledge that may be used to characterize this domain. In this case, the test base used was the glossary of the National Electrical Energy Agency (Agência Nacional de Energia Elétrica ANEEL). Similarly to other glossaries, ANEEL's glossary is a list in alphabetical order that provides a description of the meaning of words.

Soon after the list of words was generated, the hyperbolic tree was created using Hyper Tree Studio, an open-source tool based on the Hyperbolic Tree library. Figure 9 shows a hyperbolic tree whose central element is the word "Electrical Sector".

5.2 Indexing Metadata in the Hyperbolic Tree

When the Index button is clicked (Fig. 10), the editor verifies whether the metadata are validated and then stores them. After the metadata are stored, the editor creates an address that indicates where such metadata can be accessed in the future. With the creation of this address, the actual indexing process begins.

The editor uses the words typed in the field "título" of the metadata and right after that searches the nodes on the hyperbolic tree. When a node is found

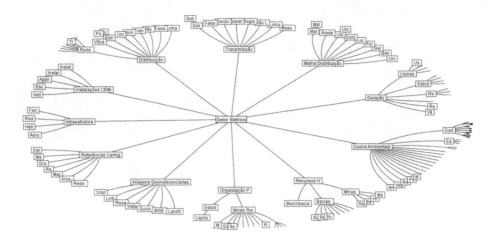

Fig. 9. Hyperbolic tree of the electric system.

Fig. 10. Panel of buttons in the edpMGB editor.

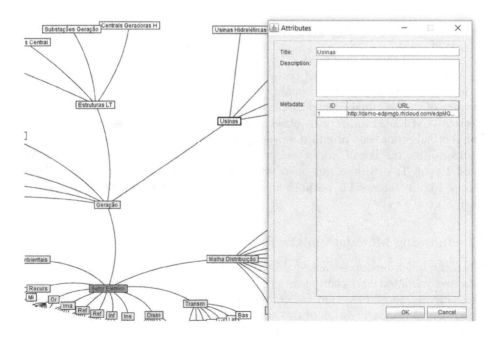

Fig. 11. Attributes box.

that has one or more words in the field "título", the editor stores within this node the address where the metadata is stored. Finally, the editor exports this tree in XML format with the metadata addresses stored. Figure 11 exemplifies this process. For example, when the node with the word "Usinas" is accessed, the user will find in the window Attributes the field Metadata, which stores the link for the metadata related to this specific term.

6   Conclusions

The present paper presented the development of a geographic metadata editor that follows the concept of Software as a Service (SaaS) and is available to any user with access to the Internet. edpMGB was developed specifically to create geospatial metadata in accordance with the Perfil de Metadados Geoespaciais do Brasil (MGB profile). Besides using software developed in his or her native language, the user has easy access to the system with no need to install it locally. edpMGB may also be used to change metadata sets created with other editors (e.g., Geonetwork).

The feature of validating whether the metadata is in accordance with the standard defined by the MGB profile helps produce higher quality, more complete and correct metadata. However, the user may save the incomplete documents, in which case the metadata receives a tag of non-conformity with the MGB profile. Since the XML validation and generation were also developed as web services, other developers will be able to remotely create applications (e.g., metadata catalog manager) that use the services implemented in the tool through SOAP, which broadens the tool's reach.

As been shown by [24], most metadata sets currently available in the INDE do not conform to the MGB profile. Therefore, this metadata editor with its XML schema conformity validation service is an important contribution to INDE's evolution.

This editor was developed in the context of the research and development (R&D) project "GeoPortal Cemig" being developed to help implement a corporate SDI for Cemig - Minas Gerais Power Company. One of the components of GeoPortal Cemig is the geospatial metadata catalog, whose metadata are documented via edpMGB, which integrates SDI-Cemig to the INDE.

It was also described the indexing module attached to the edpMGB editor that allows recovering metadata from a SDI organized within the nodes of the hyperbolic tree. The hyperbolic tree found to be adequate since it allows easy navigation in a complex network of words. The viability of the developed system was verified by developing a case study in the context of a SDI for the electricity sector, using the CEMIG database. It should be noted that the network of words used in the indexing process is a network of links that do not have a formally defined semantics, as well it was not used standardized norms to build the vocabulary. In the future, the development of a domain ontology is needed to improve the indexing process.

Finally, being free open-source software, edpMGB may also be adapted to other geospatial metadata standards and/or profiles.

As a suggestion for future works, the expansion and evolution of the editors specifications is indicated so that an ever more complete mechanism for creation, editing, and search mechanism is provided in the MGB profile. Moreover, further improving the information recovery process in a metadata catalog within the electric sector. Another extension is integrating the editor with automated extraction modules using the bounding rectangle of geospatial data and the treatment of strongly related metadata collections. Finally creating a domain ontology is key for improving the indexing process.

Acknowledgements. This project was partially funded by the Brazilian research promotion agencies Fapemig and CAPES, along with Companhia Energética de Minas Gerais - Cemig.

References

1. Armbrust, M., Fox, A., Griffith, R., Joseph, A.D., Katz, R., Konwinski, A., Lee, G., Patterson, D., Rabkin, A., Stoica, I., et al.: A view of cloud computing. Commun. ACM **53**(4), 50–58 (2010)
2. Baeza-Yates, R., Frakes, W.B.: Information Retrieval: Data Structures & Algorithms. Prentice Hall, Upper Saddle River (1992)
3. Brasil: Decreto presidencial n 6.666, de 27 de novembro de 2008 (2008). http:// www.planalto.gov.br/ccivil_03_/Ato2007-2010/2008/Decreto/D6666.htm
4. Bray, T., Paoli, J., Sperberg-McQueen, C., Maler, E., Yergeau, F., Cowan, J.: Extensible markup language (xml) 1.1-w3c recommendation 16-august-2006. World Wide Web Consortium (2006). http://www.w3.org/TR/xml11/
5. Cardoso, O.N.P.: Recuperação de informação. INFOCOMP J. Comput. Sci. **2**(1), 33–38 (2004)
6. CatMDEdit: Catmdedit opensource project (2012). http:catmdedit.sourceforge. net/index.html
7. Comissão Nacional de Cartografia - CONCAR: Perfil de metadados geoespaciais do brasil (perfil mgb) (2009). http://www.concar.ibge.gov.br/arquivo/Perfil_ MGB_Final_v1_homologado.pdf
8. Dornelles, M.A., Iescheck, A.L.: Análise da aplicabilidade da infraestrutura nacional de dados espaciais (inde) para dados vetoriais em escalas grandes. Bol. Ciênc. Geod., Sec. Artigos **19**(4), 667–686 (2013)
9. Evangelista, S.R.M.: Manual do hipereditor e hipernavegador. Embrapa Informática Agropecuária Documentos (2007)
10. Fallside, D.C., Walmsley, P.: XML schema part 0: primer second edition. W3C recommendation 16 (2004)
11. Ferneda, E.: Recuperação de informação: análise sobre a contribuição da ciência de computação para a ciência da informação. Ph.D. thesis, Escola de Comunicação e Artes Ciências Humanas, Universidade de São Paulo, São Paulo (2003)
12. Freitas, C.M.D.S., Chubachi, O.M., Luzzardi, P.R.G., Cava, R.A.: Introdução à visualização de informações. Revista de informática teórica e aplicada **8**(2), 143–158 (2001)
13. Geonetwork: GeoNetwork User Manual: Release 2.6.4. Geonetwork (2012). http://geonetwork-opensource.org/manuals/2.6.4/eng/users/GeoNetworkUserMa nual.pdf

14. Goldberg, K.H.: XML Guia prático visual. Alta Books, Rio de Janeiro (2009)
15. ISO: ISO 19115:2003. Geographic information Metadata. International Organization for Standardization (ISO) (2003)
16. ISO: ISO. ISO 19139:2007. Geographic information - Metadata: XML schema implementation. International Organization for Standardization (ISO) (2007)
17. Lamping, J., Rao, R.: The hyperbolic browser: a focus+context technique for visualizing large hierarchies. J. Vis. Lang. Comput. **7**(1), 33–55 (1996)
18. Leme, L.A.P.P.: Uma arquitetura de software para catalogação automática de dados geográficos. Ph.D. thesis, PUC-Rio, Rio de Janeiro (2006)
19. Lopes, L., Fernandes, P., Vieira, R., Fedrizzi, G.: Eχatolp-an automatic tool for term extraction from portuguese language corpora. In: Proceedings of the 4th Language & Technology Conference: Human Language Technologies as a Challenge for Computer Science and Linguistics (LTC). pp. 427–431. Faculty of Mathematics and Computer Science of Adam Mickiewicz University (2009)
20. Nakamura, E.T., Queiroz Filho, A.P.: Infraestrutura de dados espaciais: Exemplo do parque estadual de intervales-sp. Revista Brasileira de Cartografia **3**(64), 723–735 (2013)
21. Nascimento, H.A., Ferreira, C.B.: Visualização de informações-uma abordagem prática. In: XXV Congresso da Sociedade Brasileira de Computação, XXIV JAI. UNISINOS, São Leopoldo-RS (2005)
22. Nebert, D.: Developing spatial data infrastructures: The sdi cookbook v. 2.0, global spatial data infrastructure (gsdi). Open Eospatial Consortium (2004)
23. Nebert, D., Whiteside, A., Vretanos, P.: Opengis catalogue services specification (2007)
24. Pascoal, A.P., de Carvalho, R.B., de Araújo Xavier, E.M.: Materialização do perfil de metadados geoespaciais do brasil em esquema xml derivado da iso 19139. In: 16th Simpósio Brasileiro de Sensoriamento Remoto -SBSR. INPE, Foz do iguaçu -PR (2013)
25. Prado, B.R., Hayakawa, E.H., de Castilho Bertani, T., da Silva, G.B.S., Pereira, G., Shimabukuro, Y.E.: Padrões para metadados geográficos digitais: modelo iso 19115: 2003 e modelo fgdc. Revista Brasileira de Cartografia 1(62) (2010)
26. Rajabifard, A., Williamson, I.P.: Spatial data infrastructures: concept, sdi hierarchy and future directions. In: GEOMATICS'80 Conference. Tehran, Iran (2001)
27. Souza, T.B., Catarino, M.E., Santos, P.C.: Metadados: catalogando dados na internet. Transinformação 9(2) (2012)
28. Thompson, H.S., Beech, D., Maloney, M., Mendelsohn, N.: Xml schema part 1: structures second edition. W3C Recommendation, 28 October 2004. http://www.w3.org/TR
29. Vieira, S.B.: La recuperación automática de información jurídica: metodología de análisis lógico-sintáctico para la lengua portuguesa. Ph.D. thesis, Universidad Complutense de Madrid, Madrid (1994)
30. Weiss, A.: Computing in the clouds. Computing 16 (2007)
31. Yang, C., Goodchild, M., Huang, Q., Nebert, D., Raskin, R., Xu, Y., Bambacus, M., Fay, D.: Spatial cloud computing: how can the geospatial sciences use and help shape cloud computing? Int. J. Digital Earth **4**(4), 305–329 (2011)

CIAO-WPS - Utilizing Semantic Web (Web 3.0) Techniques to Assist in the Automatic Orchestration of Geospatial Processes and Datasets

Chet Bing Tan[1,2]([✉]), David A. McMeekin[1,2], Geoff West[1,2], and Simon Moncrieff[1,2]

[1] Curtin University, Bentley, WA 6102, Australia
chetbing.tan@postgrad.curtin.edu.au
https://research.haxx.net.au/
[2] Cooperative Research Centre for Spatial Information, Carlton, Australia

Abstract. Current geospatial datasets and web services are disparate, obscure and difficult to expose to the world. With the advent of geospatial processes utilizing temporal data and big data, along with datasets continually increasing in size, the problem of under-exposed datasets and web services is amplified. This paper proposes the integration of Semantic Web concepts and technologies into geospatial datasets and web services, making it possible to link these datasets and services via functionality, the inputs required and the outputs produced. To do so requires the extensive use of metadata to allow for a standardised form of description of their function. This research also visits the concept of using ontologies to store processes. A simple prototype termed CIAO-WPS (Chet's Intelligent, Automatically-Orchestrated Web Processing Services) is created as a proof of concept, using the Python programming language. The prototype seeks to reinforce ideas in regards to pathing and cost constraints, as well as explore overlooked designs.

Keywords: Semantic Web · Web 3.0 · Ontologies · Metadata · Web Processing Services · WPS · CIAO-WPS

1 Introduction

ACIL Tasman [1] determined that inefficient access to geospatial data is estimated to have reduced the direct productivity impacts in certain sectors by around \$0.5 billion, and that reductions in these inefficiencies will contribute to Australia's economic, social and environmental development goals. This is not only due to the current exposure issues of geospatial datasets and web services, but also the manual processing and workflows involved in chaining together these resources to achieve a task.

Current geospatial workflows that include geospatial web services (for example, flood prediction within a given area) are manual, requiring the chaining

© Springer International Publishing AG 2017
C. Grueau et al. (Eds.): GISTAM 2016, CCIS 741, pp. 32–48, 2017.
DOI: 10.1007/978-3-319-62618-5_3

together and intervention between the running of processes to be manually performed to ensure the relevant output is generated. The possible introduction of human error as a result of manually processing and searching the datasets and/or web services also increases the probability that the output generated is based on out-dated or even irrelevant data.

There are currently many inefficiencies with processing and using geospatial data. Yu and Liu [24] have documented these similar inefficiencies in their attempts to implement a new system that republishes real-world data as linked geo-sensor data. Janowicz [12] has also observed in their survey of semantically-enabled DSS (Decision Support Systems), that the users of said system still require a lot of work in increasing the efficiency and productivity of the intelligent system. The research outcomes of this project aligns with the conclusions of the survey.

The ability to automatically and intelligently orchestrate multiple geospatial web services to provide intelligent and useful output from a complex user query will greatly assist a geospatial analyst's efficiency, accuracy and productivity. Being able to orchestrate geospatial datasets and WPS to achieve a task will be adaptable to other fields as well. Integrating Semantic Web concepts and technologies into various non-geospatial datasets and Web Services will allow for automatic orchestration and hence, increased productivity. As an example, Kauppinen [14] has documented how semantic technologies are being integrated within Brazilian Amazon Rainforest data, that has led to increased productivity and efficiency in that area.

To advance automated and intelligent orchestration, certain features and specifications must be added to the current WPS standard to allow for machine interpretation of these Web Services, to be able to effectively and reliably determine which Web Services are appropriate for completing a given task. Critical information that paves the way for automatic orchestration are currently not defined in the WPS specification and this research aims to set an example for the addition of metadata and functions that allow for automatic orchestration as the next logical step for WPS.

This paper explores automated orchestration methods of web services and data from multiple, disparate sources; in contrast to the current widespread method of supplying all the data and services to the end user and leaving it to them to manually analyse and process the vast amounts of varying kinds of data, and determine what processing needs to be executed. Natural language processors and ontologies are proposed to build the required Artificial Intelligence to automatically chain together the resources to produce useful output for the end user.

2   Background

In the last decade, the Web has been moving towards Service-Oriented Computing architecture supporting automated use [2,11]. This architecture aims to build a network of interoperable and collaborative applications, independent of platform, called services [18,19]. The geospatial world is also moving away from the traditional desktop application paradigm to processing and accessing data on-the-fly from the Web using Web Processing Services, as outlined by Granell et al. [7]. As Web Service technology has matured in recent years, an increasing amount of geospatial content and processing capabilities are available online as Web Services [26]. These Web Services enable interoperable, distributed, and collaborative geoprocessing to significantly enhance the abilities of users to collect, analyze and derive geospatial data, information, and knowledge over the Internet [26].

Current geospatial workflows and processes rely on manual human intervention in searching for the relevant and/or required datasets and Web Services [7]. These workflows also require human analysis of the output at each stage of processing, and manual determination of which Web Processing Service to use next on the data to achieve the final required output [7]. This has been observed to lead to inefficiencies in the accuracy and currency of the data as we are relying on a human user to search for these ill-exposed datasets and Web Services. For example, a human user will tend to have a bias towards a dataset or Web Service that he/she has used before, regardless of the currency of the data or the frequency of updating of the data, a phenomenon known as the Mere-repeated-exposure paradigm [25]. Geospatial Web Services and datasets that may be vital in contributing to the final result may also be left unexposed due to current search technologies not being able to expose Web Services sufficiently. The way that Web Services are searched for is by functional and non-functional requirements as well as interactive behaviour [22], which require more than simple keyword matching, as per current search algorithms.

2.1   The Semantic Web (Web 3.0)

The Semantic Web aims to create a web of information that is machine-readable and not just human-readable [3]. This allows machines to automatically find, combine and act upon information found on the Web [20].

The objective of the Semantic Web is accomplished by integrating semantic content into web pages that helps describe the contents and context of the data in the form of *metadata* (data about data) [9]. This greatly improves the quality of the data so that a machine is able to understand what the data is for, what it can be used for and what other things are linked to it [3,10]. This allows the machine to process and use the data, instead of the current paradigm of relying on a human to interpret, process and understand data.

Ontologies are a core component of the Semantic Web, and are required by machines to be able to intelligently reason and infer data [20]. An ontology is a set of data elements within a domain that are linked together to denote the types, properties and relationships between them [4]. Ontologies contribute to resolve semantic heterogeneity by providing a shared comprehension of a given domain of interest [17]. In knowing the relationships that exist between data, search can be expanded to incorporate relationships that exist between the data as well as traditional string matching. The search is now a semantically intelligent search.

Put simply, the Semantic Web aims to create a web of knowledge and information that is both machine and human-readable [3,20]. This consequently allows the capability for machines to automatically find, combine and act upon information found on the Semantic Web.

There are currently well-defined, open standards in place for moving towards Web 3.0, where resources and ontologies are shared [21]. The advantage of this is that there is no reinvention of the wheel, however there is still significant development required in integrating Semantic Web techniques into spatial applications. To aid in the movement towards semantic spatial data manipulation, the OGC has established standards for storing, discovering and processing geospatial information [13]. Having these standards (W3C -World Wide Web Consortium and OGC - Open Geospatial Consortium) to work with creates the ability to simplify any data collected from multiple sources that are usually stored in their own unique proprietary formats, and create a standardised format of the spatial data and processes for use both in research and in industry.

2.2   The OGC WPS Standard

Web Processing Services, as their name suggests, provide services over the Internet to consumers, be it data access or processing. They are client-side platform-independent and have standardized input and output protocols so that consumers are able to utilize these Web Services [23].

It is not uncommon for a company website to be converted into an interactive, completely-automated, web-based application (such as those for stock trading, electronic commerce, on-line banking, travel agencies, etc.) [15]. It is worth noting that to achieve reliable application development, appropriate specification and verification techniques and tools are required. Systematic, formal approaches to the analysis and verification of a web service or application can deal with the problems of this particular domain by automated and trustworthy tools that also incorporate semantic aspects [15].

In the geospatial world, the OGC is the main standards body for geospatial data and technology. The standard published by the OGC in regards to Web Processing Services is currently in its second iteration (v2.1), that brings about enhanced features to improve the functionality of web services, especially in regards to asynchronous operation and error handling. However there has been little work on the orchestration of Web Services. This research requires the adding of functionality to the standard.

The OGC WPS standard sets out the groundwork for exposing features, inputs and outputs and processing of a geospatial web service [16]. Loosely defined, it simply describes the syntax and minimal requirements of a geospatial Web Processing Service.

An example of a WPS is a service that provides polygon intersection capabilities. A user provides two or more polygons as input to the WPS and receives a polygon of their intersections as output, or a NULL value if the polygons provided do not intersect. The WPS is specific as to what format and coordinate reference systems are used in the polygons that it processes.

Using and extending the WPS specification for orchestration within a workflow is not a foreign concept in the geospatial community. For example, a partially manual system of orchestrating PyWPS (A Python implementation of WPS) within Taverna has been proposed to assist in their efforts to assess modelling in urban areas [6]. This workbench for mapping business processes and workflows to chains of web services to complete geospatial tasks shows promise and feasibility.

By integrating Semantic Web concepts in the form of ontologies and metadata tags, as well as improvements and expansion of the OGC WPS standard (and possibly other open standards as seen fit), it is possible to expose all these useful datasets and Web Services that the user hasn't considered through better metadata and linking them through ontologies. The use of Semantic Web technologies allows us to look for meaning, rather than simple keyword matching. While this has been achieved to a certain degree in generic, publically-available search engines such as Google by utilizing semantic analysis algorithms [5], there has been little development in the geospatial area.

2.3   Natural Language Processing

Natural Language Processing is the process of decomposing complex queries written by the end user such that the system is able to handle the decomposed queries. For example, a NLP will be able to identify the wild cards and key phrases in a complex query such as "Chance of flooding in my area" - this is decomposed as needing a probability value (chance) for flooding in the location or region (my area).

In this research, key words and the knowledge of their type are utilized by finding similar terms within available or generated ontologies to link together services, datasets and process workflows.

2.4   Web Ontology Language and Artificial Intelligence

The latest iteration of OWL (Web Ontology Language), OWL-2, accommodates reasoning via the use of Rules, Axioms, Constraints and Ontologies within its implementation [8]. There are also specific sublanguages of OWL (in this case OWL-DL -Web Ontology Language - Description Logic) that are designed specifically for descriptive reasoning and logic. This is utilized in the scope of this research to allow for placing restrictions on the sequence of Web Services to run based on availability and information as provided by the Web Services.

A typical example of a constraint to be satisfied is that all Web Services and datasets are to be on a particular server (Fig. 1). In this situation, Process A and Process B must be located on Server X. Alternatively, a user may require the path that has the minimum response time, regardless of cost. We also use logical inference to derive potential workflows that are not explicitly present in ontologies.

This research proposes the use of OWL linked by RDF(Resource Description Framework) to provide an ontology language that is open and extensible,

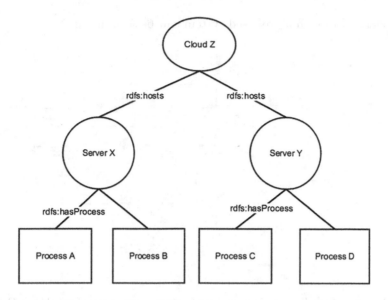

Fig. 1. An OWL visualization of fileserver constraints.

compatible with Web standards for accessibility and internationalization, and distributed across many systems. The latest iterations of OWL-2 and OWL-DL will also be explored to assist in the orchestration process as it implements rules within the language that will help with inference rules and querying.

3   Approach

To automate the orchestration process, there are several essential aspects that must be developed to assist in this regard.

Firstly, NLP (Natural Language Processing) methods require investigation to determine the meaning of a user query and semantically search for information using ontologies. In situations where a relevant ontology is not available, a classic text search may be executed and an ontology dynamically generated on-the-fly from the search results using an off-the-shelf product. In this scenario, ontologies may be used akin to a workflow, where information and process interdependencies are linked.

Secondly, Artificial Intelligence is explored that utilizes ontologies to determine what Web Services and datasets are required to generate an accurate result to the user's query. Ontologies may then be used to link together the functions, inputs and outputs of the web resources allowing the AI solution to determine which order the web services will run in to achieve an output that will satisfy the end user's query.

Finally, we identify enhancements for the OGC's open standard for WPS by adding functionality and supporting metadata to allow for automatic orchestration by the AI system developed under this research.

An overview of such a proposed system can be seen in Fig. 2.

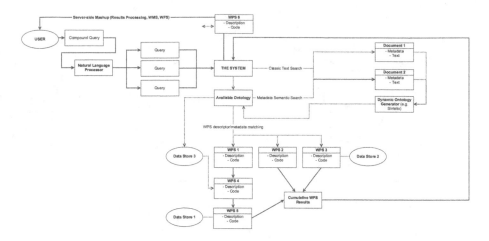

Fig. 2. A system flow diagram of how a user query is processed through CIAO-WPS.

The user firstly feeds a complex query in natural language to the system. The query is then broken down to simple, modular queries for ease of understanding using the NLP. These individual queries are then searched for in any available ontologies, and if the available ontologies are not available, the dynamic ontology generator is used to create ontologies on-the-fly using documentation and expert domain knowledge from trusted sources, obtained via classic text search from the Internet. Alternatively a system can also be put in place to query the user for additional information and details that will help complete the search.

With multiple ontologies on-hand and ready-to-use, information in regards to the user's original query can be searched within the ontologies to obtain links for, and between datasets and relevant web services. Using rules of inference and logic processing via AI, we are able to obtain a chain of Web Services and datasets in order to satisfy the user query. The sequence in which these datasets and Web Services are invoked in is also important; this is catered for by the AI. Multiple pathing and costing options can also be provided for the user to determine the most cost-efficient path to the end result (processing time or payment to use). An example of how multiple paths may lead to the same outcome is shown in Fig. 3.

The end result, however, may not always be in the most relevant form to be utilized by the user. If that is the case, we can rely on Web Services to transform the final output to a more usable form. For example, a table of rainfall predictions is not as readily-usable to a human user in comparison to a percentage chance of rain in an area, drawn on a map using a WFS (Web Feature Service) or WMS (Web Map Service) derived from the table data to show different rainfall chances in different areas of the map via heat layers.

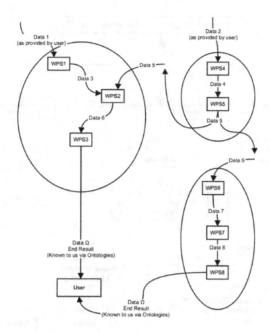

Fig. 3. Different paths for the same outcome.

4   Use Case Scenario 1 - Flood Evacuation Model

A typical scenario that shows how CIAO-WPS may be utilized is shown in Fig. 4. The following paragraphs will guide the reader through the diagram.

In this example, the user feeds a complex, natural-language query into the system. NLP breaks the query down into three separate sections. "What are the chances", "of flooding" and "in my area".

Based on "in my area", CIAO-WPS realizes the statement implies location of the user and therefore the first step would be to obtain the location of the user as this will provide very important context for the other queries. This could use simple GPS on the user's smart phone/device.

With an ontology such as the SWEET (Semantic Web for Earth and Environmental Technology) ontology readily available to CIAO-WPS, the system looks up "flooding" within the ontology, that will give links to elements such as "rain", "water", "severe weather condition". These elements will further give links to information such as links to BOM (Bureau of Meteorlogy) severe weather warning services or flood prediction Web Services, for example. As for the "in my area" section, the location of the user becomes extremely important at this point as CIAO-WPS will focus on the vicinity of the user.

"What are the chances" prompts CIAO-WPS that this is most likely a percentage/likelihood calculation. This contextual information is fed into the search for related Web Services and datasets.

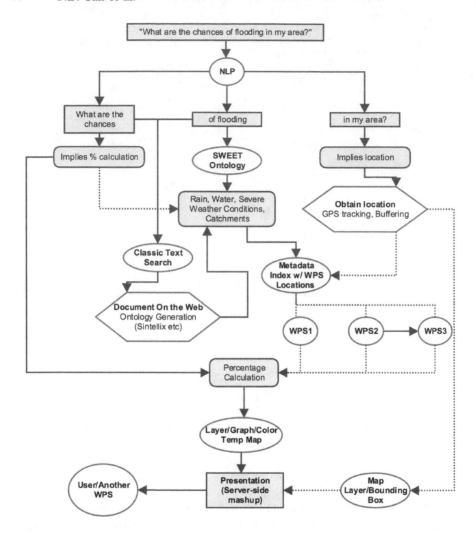

Fig. 4. A sample use case scenario of CIAO-WPS.

With the query broken down, CIAO-WPS will attempt to use traditional text-search tools or further delve into more specialized ontologies to obtain more information and methodologies to obtain the relevant Web Services and datasets. This in turn could reveal more critical information such as workflows that provide a more authoritative source of guidance as to what Web Services and what datasets to access in what order.

Using AI and OWL-DL, CIAO-WPS attempts to obtain a path that fits in with user-defined restrictions and will fulfill the user's query. If a section or piece of information along the path is unavailable, CIAO-WPS will notify the user and fallback into a semi-supervised mode, in which the user may provide the needed information or direct which path CIAO-WPS should take.

In this case, we assume that all information is obtained successfully within the user's set parameters. The results are accumulated by CIAO-WPS. However what the system has obtained is simply polygons in the vicinity with a percentage chance of flooding attached.

This would not be useful to the user and therefore CIAO-WPS draws the polygons on a map, while coloring in the chances of flooding with different intensities using a heat scale. This is much more useful to a human user than presenting the user with a table of numbers and coordinates for the edges of the polygon.

5   Use Case Scenario 2 - Coral Reef Approximation

5.1   The Problem

In Australia, Geoscience Australia (GA) have a DataCube project where they make available to the public a large amount of their resources and datasets (in the form of WMS and WFS). They currently face issues with making all the available datasets useful to the general population. A specific, small-scale application of CIAO-WPS can be found here in this scenario.

A common question encountered by the organization is"Show me coral reefs within x amount of kms from y". This kind of enquiry has applications with protected environments within Australia. For example, fishing companies require knowledge as to where they are allowed to fish so as to not damage precious coral reefs. Another application of such a query would be for offshore mining companies to plan out drilling explorations.

To answer this query, Geoscience Australia uses two of their datasets - bathymetry (depth) and backscatter (hardness) of Australia's ocean beds. Generally, a good indication that a plot of seabed is a coral reef is an area that is shallow and hard. This provides a rough, but useful approximation as to where a coral reef may be found.

5.2   Background

To be able to see the application of CIAO-WPS in such an application, we first need to understand what bathymetry and backscatter are. A basic, brief description of both terms is sufficient for the reader to understand how they may used to get an approximation of coral reef locations. Therefore, this paper will not delve into too much detail as to how the datasets are obtained; rather, this paper focusses on describing what the datasets are.

Bathymetry refers to the depth of the seabed at a specified location. The unit of measurement for bathymetry is metres (below sea level). A higher reading of metres below sea level simply represents a deeper sea bed. Bathymetry information is useful as different sea bed levels could attribute themselves to a lot more information when paired with other datasets. For example, certain photogrammetry datasets (especially to do with chlorophyll) combined with bathymetry datasets may lend themselves to revealing more information about the flora/ecology of a certain area.

Backscatter represents the hardness of a surface. Backscatter data is obtained from multi-beam echosounders that scan the seafloor. A sonar is sent to the depths and the reflection of the wave is measured. The unit of measurement is dB (decibels) - a stronger, higher dB reading represents a harder surface. The reason behind this is that a more solid surface will have a higher reflectivity of sonar, in contrast to a soft surface that would"absorb" most of the wave, resulting in a weaker feedback.

5.3   Potential Application of CIAO-WPS

CIAO-WPS was designed to implement automated orchestration. In this scenario, we already have the knowledge as to what datasets, what constraints and what operations are required to achieve the result we want to obtain a useful answer.

In this coral reef use case scenario, we know that we require the two datasets - bathymetry and backscatter. Constraints are applied to these datasets and they are required to be intersected to produce a good approximation of coral reef locations.

Geoscience Australia provides the two datasets as WMS. These can be exposed as web services easily (built into the standard), or in the worst case

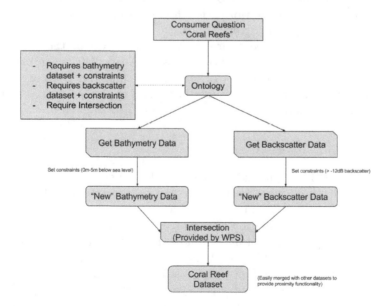

Fig. 5. Coral Reef Approximation concept.

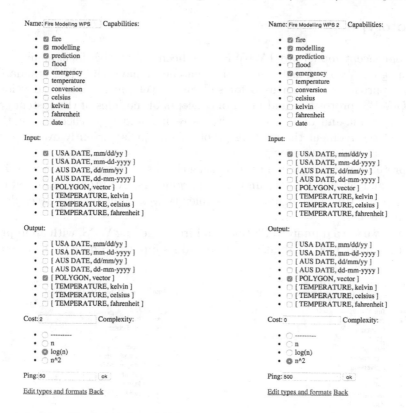

Fig. 6. Dummy Fire Modelling WPS 1.   **Fig. 7.** Dummy Fire Modelling WPS 2.

scenario, their inputs and outputs mapped to a process ontology, whereby an ontology holds information as to what tasks and what datasets must be accessed and performed to achieve a process flow. As this is a commonly asked question, it is possible that the organization holds a process ontology of tasks, each of which contain all the steps and information as to how to complete such a task.

For example, a domain expert or geospatial analyst will be able to input into our process workflow ontology the constraints for the two datasets (between 0 m and 5 m below sea level and > -12dB backscatter). These two datasets are then intersected to produce the approximation for coral reefs, and then this resulting dataset can then be bounding box-ed to other datasets, such as coast lines (for proximity to coast lines) or even further intersected to identify coral reefs in protected marine areas (another dataset).

The process flow for CIAO-WPS integration into the task can be visualized in Fig. 5.

6   Prototype

A proof-of-concept model of CIAO-WPS has been built, with focus mainly on the chaining of WPS via metadata added outside of the OGC WPS standard. An Agile approach has been used for software development used in developing CIAO-WPS, prototypes and proof-of-concepts of modules of the system are continuously built and discarded to serve as reality and logic checkers for the research. At the moment the working proof-of-concept is currently available at http://research.haxx.net.au/cwps.

The prototype was created in Python using classes to serve as "dummy" WPS with input and output formats and their parameters. Additional information such as response time and algorithmic complexity were added as well to explore these ideas further.

In Fig. 6, we see a dummy WPS that is a Fire Modelling WPS, with appropriate metadata in its GetCapabilities section and its input (US date in the format

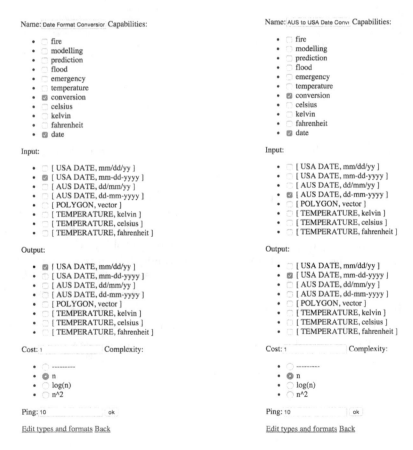

Fig. 8. Dummy date format conversion WPS.

Fig. 9. Dummy AUS to US date conversion WPS.

mm/dd/yyyy). It outputs a polygon in vector format of the predicted fire area. This WPS has metadata that states that it has a high cost but a quick response time and a relatively simple algorithmic complexity.

In Fig. 7, we see another dummy WPS that also performs fire predictions, with appropriate metadata in its GetCapabilities section and its input (US date in the format mm/dd/yyyy), similar to the first fire modelling WPS. It outputs a polygon in vector format of the predicted fire area. The metadata of this version states that it has no cost but a slower response time and a relatively more complex algorithmic complexity.

In Fig. 8, we see a dummy WPS that is a date format conversion WPS, with appropriate metadata in its GetCapabilities section and its input (American date in the format mm-dd-yyyy). It outputs an American date in the format mm/dd/yyyy.

In Fig. 9, we see a dummy WPS that is a regional date conversion WPS, with appropriate metadata in its GetCapabilities section and its input (Australian date in the format dd-mm-yyyy). It outputs an American date in the format mm-dd-yyyy.

The interface for this proof-of-concept can be observed in Fig. 10, where the user wants to do "fire modelling", but only has an Australian date in the format dd-mm-yyyy as input.

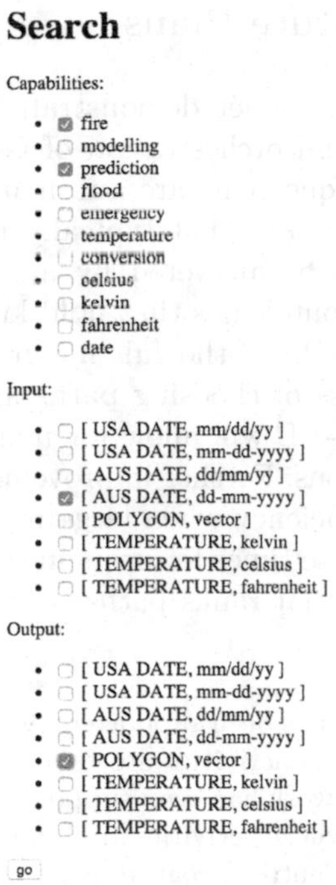

Search

Capabilities:

- ☑ fire
- ☐ modelling
- ☑ prediction
- ☐ flood
- ☐ emergency
- ☐ temperature
- ☐ conversion
- ☐ celsius
- ☐ kelvin
- ☐ fahrenheit
- ☐ date

Input:

- ☐ [USA DATE, mm/dd/yy]
- ☐ [USA DATE, mm-dd-yyyy]
- ☐ [AUS DATE, dd/mm/yy]
- ☑ [AUS DATE, dd-mm-yyyy]
- ☐ [POLYGON, vector]
- ☐ [TEMPERATURE, kelvin]
- ☐ [TEMPERATURE, celsius]
- ☐ [TEMPERATURE, fahrenheit]

Output:

- ☐ [USA DATE, mm/dd/yy]
- ☐ [USA DATE, mm-dd-yyyy]
- ☐ [AUS DATE, dd/mm/yy]
- ☐ [AUS DATE, dd-mm-yyyy]
- ☑ [POLYGON, vector]
- ☐ [TEMPERATURE, kelvin]
- ☐ [TEMPERATURE, celsius]
- ☐ [TEMPERATURE, fahrenheit]

go

Fig. 10. CIAO-WPS proof-of-concept search interface.

Path

Shortest Path

Path [<WPS: AUS to USA Date Conversion WPS>, <WPS: Date Format Conversion WPS>, <WPS: Fire Modelling WPS>] with cost 4, complexity $O(n + n + \log(n))$, and ping 70.

Simplest Path

Path [<WPS: AUS to USA Date Conversion WPS>, <WPS: Date Format Conversion WPS>, <WPS: Fire Modelling WPS>] with cost 4, complexity $O(n + n + \log(n))$, and ping 70.

Cheapest Path

Path [<WPS: AUS to USA Date Conversion WPS>, <WPS: Date Format Conversion WPS>, <WPS: Fire Modelling WPS 2>] with cost 2, complexity $O(n + n + n^2)$, and ping 520.

Lowest Ping

Path [<WPS: AUS to USA Date Conversion WPS>, <WPS: Date Format Conversion WPS>, <WPS: Fire Modelling WPS>] with cost 4, complexity $O(n + n + \log(n))$, and ping 70.

All Paths

- Path [<WPS: AUS to USA Date Conversion WPS>, <WPS: Date Format Conversion WPS>, <WPS: Fire Modelling WPS>] with cost 4, complexity $O(n + n + \log(n))$, and ping 70.
- Path [<WPS: AUS to USA Date Conversion WPS>, <WPS: Date Format Conversion WPS>, <WPS: Fire Modelling WPS 2>] with cost 2, complexity $O(n + n + n^2)$, and ping 520.

Fig. 11. CIAO-WPS proof-of-concept results.

Finally in Fig. 11, we have the results displayed. CIAO-WPS does a custom-tuned BFS (Breadth-First Search) to explore successful paths and choose the path that satisfies the user's constraints. At this stage, the prototype will reveal multiple options for shortest, simplest, cheapest and most responsive paths. This idea translates to a semi-supervised operation in the final system.

7   Conclusion and Future Plans

The use cases outlined in this paper demonstrate the ability for CIAO-WPS to assist in the automation and orchestration of Geospatial Web Services and datasets. Commonly-queried questions are automated to increase productivity and efficiency, while urgent queries that typically require a geospatial analyst and/or a domain expert can be answered by an automated system, making informed choices by utilizing ontologies that hold knowldge.

Future plans for CIAO-WPS is the full integration and use of ontologies in the decision-making process of choosing paths and the order in which Web Services are run. Using OWL-DL will allow for more complex constraints such as file server location restrictions. Further improvements to the search algorithm are also planned for better efficiency when large numbers of WPS are involved. This ensures consistent response times for users in time-demanding applications. Finally, a learning algorithm that ranks paths based on past successful queries will also be explored.

Acknowledgement. The work has been supported by the Cooperative Research Centre for Spatial Information, whose activities are funded by the Business Cooperative Research Centres Program. The work has been supported by the Cooperative Research Centre for Spatial Information, whose activities are funded by the Australian Commonwealth's Cooperative Research Centres Programme.

References

1. ACIL Tasman: The Value of Spatial Information - the impact of modern spatial information technologies on the Australian economy. Technical report (2008). http://www.crcsi.com.au/assets/Resources/7d60411d-0ab9-45be-8d48-ef8dab5abd4a.pdf
2. Ameller, D., Burgués, X., Collell, O., Costal, D., Franch, X., Papazoglou, M.P.: Development of service-oriented architectures using model-driven development: a mapping study. Inf. Softw. Technol. **62**, 42–66 (2015)
3. Berners-Lee, T., Hendler, J., Lassila, O.: The Semantic Web. Sci. Am. **284**(5), 28–37 (2001)
4. Beydoun, G., Low, G., García-Sánchez, F., Valencia-García, R., Martínez-Béjar, R.: Identification of ontologies to support information systems development. Inf. Syst. **46**, 45–60 (2014)
5. Cilibrasi, R.L., Vitányi, P.M.B.: The Google similarity distance. IEEE Trans. Knowl. Data Eng. **19**(3), 370–383 (2007)
6. De Jesus, J., Walker, P., Grant, M., Groom, S.: WPS orchestration using the Taverna workbench: the eScience approach. Comput. Geosci. **47**, 75–86 (2012)
7. Granell, C., Díaz, L., Tamayo, A., Huerta, J.: Assessment of OGC web processing services for REST principles. Int. J. Data Min. Model. Manage. **6**(4), 391–412 (2012). (Special Issue on Spatial Information Modelling, Management and Mining)
8. Grau, B.C., Horrocks, I., Motik, B., Parsia, B., Patel-Schneider, P., Sattler, U.: OWL 2: the next step for OWL. Web Semant. Sci. Serv. Agents World Wide Web **6**(4), 309–322 (2008)
9. Handschuh, S., Staab, S.: CREAM: CREAting metadata for the Semantic Web. Comput. Netw. **42**(5), 579–598 (2003)
10. Harth, A.: An integration site for Semantic Web metadata. Web Semant. Sci. Serv. Agents World Wide Web **1**(2), 229–234 (2004)
11. Huhns, M.N., Singh, M.P.: Service-oriented computing: key concepts and principles service-oriented computing: key concepts and principles. IEEE Internet Comput. **9**(1), 75–81 (2005). http://scholarcommons.sc.edu/csce_facpub, http://ieeexplore. ieee.org/servlet/opac?punumber=4236
12. Janowicz, K., Blomqvist, E.: The use of Semantic Web technologies for decision support - a survey. Semant. Web J. **5**(3), 177–201 (2012)
13. Janowicz, K., Schade, S., Keßler, C., Maué, P., Stasch, C.: Semantic enablement for spatial data infrastructures. Trans. GIS **14**(2), 111–129 (2010)
14. Kauppinen, T., Mira De Espindola, G., Jones, J., Sánchez, A., Gräler, B., Bartoschek, T.: Linked Brazilian Amazon rainforest data. Semant. Web **5**(2), 151–155 (2014)
15. Kovács, L., Kutsia, T.: Special issue on Automated Specification and Verification of Web Systems. J. Appl. Logic **10**(1), 1 (2012)
16. Lopez-Pellicer, F.J., Rentería-Agualimpia, W., Bé, R.N., Muro-Medrano, P.R., Zarazaga-Soria, F.J.: Availability of the OGC geoprocessing standard: March 2011 reality check. Comput. Geosci. **47**, 13–19 (2012)
17. Nacer, H., Aissani, D.: Semantic Web services: standards, applications, challenges and solutions. J. Netw. Comput. Appl. **44**, 134–151 (2014)
18. Papazoglou, M., Georgakopoulos, D.: Introduction: service-oriented computing. Commun. ACM - Service-Oriented Comput. **46**(10), 24–28 (2003)
19. Pugliese, R., Tiezzi, F.: A calculus for orchestration of web services. J. Appl. Logic **10**(1), 2–31 (2012). www.elsevier.com/locate/jal

20. Pulido, J.R.G., Ruiz, M.A.G., Herrera, R., Cabello, E., Legrand, S., Elliman, D.: Ontology languages for the semantic web: a never completely updated review. Knowl.-Based Syst. **19**(7), 489–497 (2006). www.elsevier.com/locate/knosys
21. World Wide Web Consortium: W3C Semantic Web FAQ (2001). http://www.w3.org/2001/sw/SW-FAQ
22. Wu, Z.: Service discovery. In: Service Computing - Concept, Method and Technology, Chap. 4, pp. 79–104 (2015)
23. Wu, Z.: Service-oriented architecture and web services. In: Service Computing - Concept, Method and Technology, Chap. 2, pp. 17–42 (2015)
24. Yu, L., Liu, Y.: Using linked data in a heterogeneous sensor web: challenges, experiments and lessons learned. Int. J. Digital Earth **8**(1), 17–37 (2013). http://www.scopus.com/inward/record.url?eid=2-s2.0-84920191714&partnerID=tZOtx3y1
25. Zajonc, R.B.: Mere exposure: a gateway to the subliminal. Curr. Dir. Psychol. Sci. **10**(6), 224–228 (2001)
26. Zhao, P., Lu, F., Foerster, T.: Towards a geoprocessing web. Comput. Geosci. **47**, 1–2 (2012)

Investigation of Geospatially Enabled, Social Media Generated Structure Occupancy Curves in Commercial Structures

Samuel Lee Toepke[✉]

Private Engineering Firm, Washington DC, USA
samueltoepke@gmail.com
http://www.samueltoepke.com

Abstract. Spatiotemporal human-use estimations, via occupancy curves of a high-traffic commercial structure, have been shown to be attainable using publicly available social media data. The data is crowd sourced, geospatially enabled, and gathered from open web services using a commercially available, enterprise cloud architecture. After data processing, an interested individual can view a graph displaying population over a twenty four hour period for a specific building, with this work focusing on several structures in downtown San Jose, CA, USA. New structure data is explored to bolster previous findings, structure curves are compared to Google Popular Times charts, and further discussion includes limitations of this method and the benefit of error estimation.

Keywords: Population estimation · Structure occupancy curve · Social media · Geofencing · Enterprise architecture · Volunteered geographic data

1 Introduction

Population distribution estimation continues to be a critical problem, with areas of impact including emergency response, crisis management, energy use projection, and urban planning [18]. A subset of the population distribution problem includes discerning the spatiotemporal population inside of a single building during occupied hours.

This task is currently completed using several methods, all require high installation/ongoing costs and maintenance:

– Electronic access-control, which limits the use of the structure to previously cleared users, requiring the use of an identification card or token.
– Staffed access-control, which requires a security team and all the requisite training, administration and scheduling.
– Measurement of consumables, such as Internet IP addresses and/or power use.
– Networked acoustic and/or infrared sensors.

In modern society, smart phones and other Internet connected devices have become pervasive; iPhones, Androids and Blackberrys are carried by almost two thirds of the American adult population [32] and are connected to the Internet through Wi-Fi and/or

© Springer International Publishing AG 2017
C. Grueau et al. (Eds.): GISTAM 2016, CCIS 741, pp. 49–61, 2017.
DOI: 10.1007/978-3-319-62618-5_4

wireless carriers. These devices generally have GPS functionality, allowing the user to make geospatially enabled posts on social media sites such as Twitter and Instagram.

Twitter and Instagram expose a public application programming interface (API) that allows interested users to access posts using web services and a compatible programming language. If these posts have an associated location and the post density is high enough, this information can be used for population estimation [2]. Geofencing is the practice of filtering the geospatially enabled posts to a specified geographic boundary [24]; if the boundary is the perimeter of a structure, the resulting posts can provide a useful basis for occupancy curve estimation.

This investigation is an expanded version of [37], and shows a use case of downtown San Jose, CA, USA. Several buildings of different size are investigated; implementation notes are discussed, and results are presented.

2   Background

Research into population estimation is currently moving forward in two complementary directions: with data sourced from traditional methods, and with information harvested from electronic sources.

The first approach uses a combination of rigorously collected information to generate estimations. LandScan USA [4] is a population estimation product produced by Oak Ridge National Laboratory, USA. LandScan uses a fusion of census data, administrative boundaries, raster/vector data, and high resolution images to generate a dasymetric map [23] that has an approximate resolution of 1 km^2 [28]. Urban Atlas, a similar product from the European Commission, generates its population estimation data from census tracts, horizontal soil sealing, land-use/cover maps, commune boundaries, etc. [3] Using data disaggregation and weighting, polygon maps for specific areas are created.

The second track occurs through the active investigation of volunteered geographic information from social media services such as Twitter, Instagram, Facebook, Foursquare, Panaramio, etc. The user generated content from these services provides a wide variety of data inexpensively, that can be queried programmatically, while leveraging the idea of using humans as mobile sensors [2,20].

Recent work into mining of social media data includes population estimation for the purpose of emergency response [36], modelling population at risk in active volcanic areas [10,11], tracing the German centennial flood [12], and creating high resolution mapping of special events [31].

On July 28, 2015, Google introduced a feature named Popular Times (PT) in their search engine. Instead of using publicly available social media posts, location data from cooperating Android devices is used to generate occupancy curves [26]. While the source data is not publicly available for general research, and an end user cannot currently view all structures, this is a convenient way to generate/view occupancy curves.

The traditional model is based on broad and well researched data; though it is slow to deploy, it has a low spatiotemporal resolution and is expensive to implement. The social media model has a rapid turnaround, and can be very dense; but is only tenable in populated areas that have a high level of tech adoption, and a user base with a propensity to generate posts. Ideally, the fusion of both methods can provide a more flexible and inexpensive population distribution model. E.g., the weekday/daytime distribution found in this study could contribute to more precise workplace zone data [21].

Modern social media population estimations are mainly focused in the emergency response and crisis management arenas; but can also be applied in other areas. Individual location data throughout a day is critical to structure occupancy planning, and social media can be of use. Currently, structure occupancy curves can be obtained through simulations [27], direct sampling of building use [7], and/or measuring of consumables e.g. Internet/power/water. The methods are effective, but can be expensive; requiring sensor suites, on-site personnel, and access to building statistics. The convenience of occupancy estimations from public sources is also pertinent to interested third parties e.g. an Internet service provider, who would have no expectation of attaining this data through rote channels.

3   Architecture

The data used is gleaned from a previous investigation [36] and consists of geospatially enabled posts from Twitter and Instagram occurring from 05.16.2014 00:00:00 (GMT) to 12.31.2014 23:59:59 (GMT). Publicly available web service APIs were used to download the data in a JavaScript Object Notation (JSON) format.

Data purchased from GNIP [14] was briefly considered for this investigation. Using the same geographic bounding box as the publicly available data, a dense amount of historical results was found, averaging approximately 13,000 Tweets per month. However, the cost for the data was out of scope at this stage of investigation [13].

The publicly available posts were collected using a Java Platform Enterprise Edition (J2EE) [25] application deployed to the cloud. Google App Engine (GAE) [15] was the infrastructure selected, and full use was made of the datastore, user access and job scheduling APIs. The infrastructure was chosen for convenience, low cost and high availability; though the Twitter and Instagram APIs are web service based, any compatible programming language or enterprise architecture could have been leveraged.

Once collected, the data was inserted into an open source PostgreSQL database installed with the PostGIS extension. The extension allows geospatial queries on the data; primarily used to return records around a geofenced area.

Several buildings of interest were identified, and geospatial queries were created to obtain the necessary data. E.g., to query the Tweets from around the San Jose Convention Center, the following query was used:

```
SELECT * FROM twitter_data
WHERE timestamp < '2014-12-31
23:59:59 +00' AND timestamp > '2014-05-
16 00:00:00 +00'
AND
ST_contains(ST_MakePolygon(ST_GeomFromT
ext('LINESTRING(-121.890406 37.329801,
-121.889982 37.329170, -121.890669
```

```
37.328850, -121.889794 37.327613, -
121.886822 37.329409, -121.887664
37.330202, -121.888528 37.329865, -
121.888936 37.330428, -121.890406
37.329801)',3857)), "location");
```

The 'timestamp' portion of the query limits results to the target dates, and the 'ST_contains' portion creates a geofence, and returns the posts from the interior.

When creating the geospatial queries, projection selection is of critical importance. The social media posts are available in the EPSG:3857 projected coordinate system [8]. It is essential that the 'location' field for each record in the database as well as the previously shown query are also in the same projection. Ignoring this detail can return results that appear correct, but are inaccurate.

Choosing points for the geofence is most easily done using a point selector that is coded in Google Maps, or any other EPSG:3857 projection. Zooming in as far as possible, while making sure to obtain as much of the building as possible will give the best results.

GPS accuracy also needs to be considered when selecting the query border. Horizontal GPS precision is currently claimed to be approximately 4 meters RMS [16]. Readings can be ameliorated by having a clear view of the sky, having many satellites locked to the device, etc. The GPS reading can also be degraded by electromagnetic interference, adjacent buildings, quality of device/antenna, etc. For a standalone structure like the SJSU Event Center, this is less of an issue. When attempting to attain occupancy curves of a structure that has highly trafficked structures on adjacent walls or is vertically occluded, estimations can be negatively affected. This effect can be mitigated by check-in functionality, which is currently implemented by specific social media services. E.g. if a user is inside a coffee shop, they can use wireless Internet to report their presence. No GPS functionality is required, and their spatiotemporal use of that space is recorded effectively.

San Jose, CA exists in a temperate environment, such that outdoor space is integral to a building's usable area. Judicious selection of the geofence will allow for a more precise view of building use, even if the use area is outdoors. E.g. the SJSU student union has an outdoor seating area southwest of the building. This area was included in the geofence because it is maintained by the building, and patrons will be utilizing this space. On the other hand, it is difficult to ascertain the border for the Performing Arts Center. The building is surrounded by a plaza that allows for congregation before/after performances, but picking the actual border is not exact. Firsthand knowledge of the area, as well as observed use-patterns can be beneficial in geofence selection.

The structures for this investigation were picked based on expectation of a population utilizing social media services, as well as being non-residential. Residential buildings offer difficulty, as they are never officially open/closed, and sleeping residents will not be posting, thus creating a skewed population expectation. Also, residential buildings do not have the population density required to provide an adequate population estimation.

The structures selected include:

- San Jose Convention Center: the primary convention center in the city of San Jose, with over 500,000 sq. ft. of event space and a convenient downtown location [6].
- San Jose State University (SJSU) Student Union: a building hosting services for students of the university [34].
- SJSU Event Center: capable of holding 7,000 individuals, this space hosts the FIRST Robotics Competition in Silicon Valley [9].
- SJSU Dr. Martin Luther King, Jr. Library: the main campus library for the university.
- San Jose Center for the Performing Arts: a large theatre in downtown [5].
- San Jose City Hall: main civic administration building for the city [29].
- The Tech Museum of Innovation: a museum in downtown San Jose with an IMAX theater and exhibits focusing on energy efficiency, exploration and genetics [35].
- SJSU Dining Commons: food hall for students at the University.
- San Jose Children's Discovery Museum: a space dedicated to education [1].
- La Victoria Taqueria: a popular restaurant known for inexpensive Mexican food [19].
- San Jose Repertory Theatre: a residential professional theatre company [30].
- San Jose Convention Center, South Hall: a standalone 80,000 sq. ft. exhibit space [33].
- Naglee Park Garage: a neighborhood bistro serving new American cuisine with a large outdoor sitting area.

Once the queries have been created, and the buildings selected, the occupancy curves were created with the following pseudocode.

```
for each structure
    for each day of week (SMTWTFS)
        get Twitter/Instagram count for each hour
        get number of unique days (e.g. Mondays) in data sample
        average count for each hour of each day by number of specific
        weekdays display resulting data in a JavaScript chart
```

The pseudocode was implemented in Java, and interacts directly with the Post-greSQL database using Java Database Connectivity. The results for each structure are displayed in a website utilizing Highcharts, an interactive charts plug-in for JavaScript [17].

This investigation uses a snapshot of previously collected data as a standalone prototype. When the curve generation code is running along with the Twitter/Instagram collection code, a regularly updating aggregate estimation for each structure can be maintained; resulting in an always up-to-date, pseudo real-time estimation.

4   Results/Observations

The code was run on the thirteen structures; Table 1 shows the counts of each set of results.

For each structure, a Google Map web page was made, with each social media post being represented as a balloon, to verify all posts are within the expected borders. An example of this map can be seen in Fig. 1.

Table 1. Total Tweet/Instagram Counts per Structure, 05.16.2014 to 12.31.2014.

Venue	Tweets	Instagrams	Total
Convention Center, no S. Hall	5361	4754	10115
SJSU Student Union	1613	2205	3818
SJSU Event Center	1959	1529	3488
SJSU Library	1597	766	2363
Center for the Performing Arts	655	788	1443
City Hall	602	662	1264
The Tech Museum	595	653	1248
SJSU Dining Commons	1066	79	1145
Children's Discovery Museum	133	441	574
La Victoria Taqueria	158	90	248
Repertory Theatre	97	142	239
Convention Center, S. Hall	79	95	174
Naglee Park Garage	21	54	75

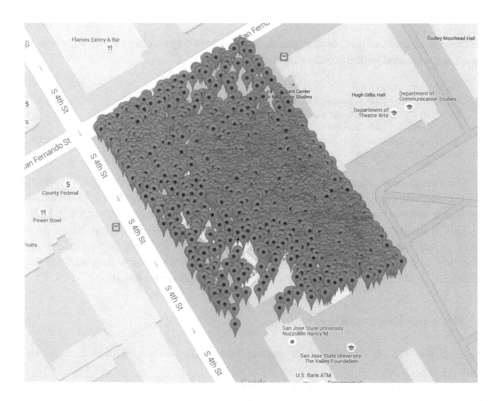

Fig. 1. Results of query on SJSU Library.

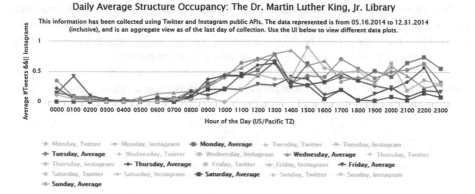

Fig. 2. Structure Curves, SJSU Library.

Also for each structure, an interactive JavaScript web page was created that allows the user to view Twitter/Instagram posts individually, or as an average, for each day. The average occupancy for the SJSU library is shown in Fig. 2.

Observation of the resulting charts of each structure shows distinct patterns up until the SJSU Dining Commons. For the remaining structures, it is difficult to tell whether the results are noise, or actual population estimation results. Though even the Naglee Park Garage, a popular restaurant, which has the lowest count of social media posts, shows posts around what would be considered lunch/dinner time.

Each graph can also be viewed as a bar graph. In Fig. 3, a reading from the San Jose Convention Center on a Thursday is shown. There is a morning rush, a lull after lunch, with the rest of the afternoon becoming slightly stronger before tapering off.

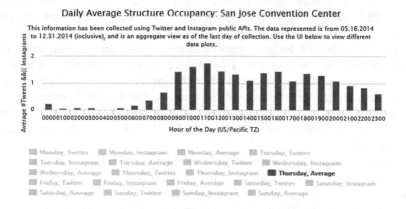

Fig. 3. Structure Curve, SJSU Convention Center, Thursday, Average.

It is of note that results exist outside the regular building hours, for each structure. This is likely a result of the aforementioned GPS horizontal error in the social media posts. While the results could be from individuals who are inside after hours, they likely come from people who are walking by, or standing near the buildings.

Figure 4 shows a full representation of each structure, for a Thursday. One can see a clear ebb and flow of posts throughout the day, as well as a drastic drop when the structures are meant to be closed.

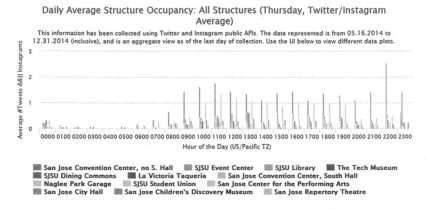

Fig. 4. Structure Curves, All Structures, Thursday, Average.

The Google PT curves offer an opportunity for comparison/contrast, for implemented buildings. The data is sourced from Android users who have chosen to share their locations [26]; and depends on Android pervasiveness as well as user-opt in. Unfortunately, PT is only implemented for four of the buildings in this investigation: The SJSU Library, The Tech Museum, La Victoria, and the Children's Discovery Museum.

It would be beneficial for the purposes of this investigation to have available as many PT charts as possible. Alas, the PT website claims the chart will only appear for businesses that have sufficient data, and whose business hours are listed in Google. As previously mentioned, the structures with less results than the SJSU Dining Commons appear to show noise, without obvious patterns. Only the The Tech Museum and the SJSU Library have enough data to be compared with their PT charts, with Thursday chosen arbitrarily.

The structure occupancy curve for The Tech Museum, with 1248 tweets/Instagrams, can be seen in Fig. 5. The PT chart, collected on 07.04.2016 can be seen in Fig. 6.

The structure occupancy curve for The SJSU Library, with 2363 tweets/Instagrams, can be seen in Fig. 7. The PT chart, collected on 07.04.2016 can be seen in Fig. 8.

An approximately similar increase to peak-use is demonstrable in both structures. While the generated occupancy curves show a large dip immediately post-peak, the PT charts show a smooth downward transition. There are two working hypotheses for this discrepancy:

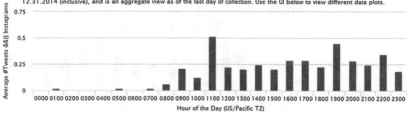

Fig. 5. Structure Curve, Thursday, Average, The Tech Museum.

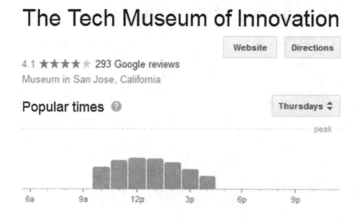

Fig. 6. PT Chart, Thursday, The Tech Museum.

- PT is based on a check-in model, while this investigation is based on post-volume aggregation. It is likely that a contributor does not consistently post tweets/Instagrams throughout the entire time they are in a structure. This lack of regular posting can cause the population to be under-estimated, which is a drawback of this method.
- Google could be applying standard filtering algorithms including smoothing and outlier-removal, which would minimize sudden drops/jumps in the charts.

Also of note is that PT doesn't show data outside of open hours, this data is likely truncated to prevent confusion for the end user. Further work would greatly benefit from more data density in the structure curves, objective building occupancy data for comparison, and transparent insight into the PT algorithms.

Comparing the structure occupancy curves with known patterns is the most effective way to validate this estimation method. Unfortunately, traditional population density generation methods do not have the spatiotemporal precision necessary to generate an adequate comparison. Disaggregation methods used in NDPop [10], and the Urban Atlas polygons [3] make progress towards combining traditional sources and honing this precision.

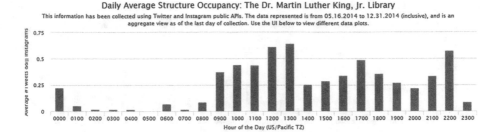

Daily Average Structure Occupancy: The Dr. Martin Luther King, Jr. Library

This information has been collected using Twitter and Instagram public APIs. The data represented is from 05.16.2014 to 12.31.2014 (inclusive), and is an aggregate view as of the last day of collection. Use the UI below to view different data plots.

Fig. 7. Structure Curve, Thursday, Average, The SJSU Library.

Dr. Martin Luther King, Jr. Library

4.5 ★★★★ 113 Google reviews

Library

Popular times ❓

Fig. 8. PT Chart, Thursday, The SJSU Library.

Another method for validation would be partnering with buildings that currently implement an active access-control scheme. Assuming the measures are implemented effectively, the data would provide a compelling comparison against the curve estimations.

The full results, including Google Map representations of the data queries, as well as the JavaScript charts for each structure, with curves for each day of the week, can be obtained by emailing the author.

5   Follow-On Work

This is a cursory investigation, to show proof of concept for structure occupancy estimation. The current algorithm will benefit from the following.

– Denser data and/or a longer collection period. Full-stream data, purchased from GNIP would give a more realistic representation of the social media posts, as only a subset is publicly available through the official APIs. The longer collection period will give a more robust average, resilient to rapid and/or irregular population swings.

- Integration of method with other novel population distribution methods e.g. NDPop, and as part of a data source for a more precise spatiotemporal model [22].
- Social media post processing. Identifying/removing cross-posts from different services will allow the curves to more accurately represent a human population. Limiting the number of posts per user in a specified time frame will also prevent skewing of the estimation. Also, removal of any nefarious attempts to skew the data. E.g. several individuals continuously posting at a restaurant to make it appear more popular, thus driving more business.
- Sensor filtering. Integrating filtering approaches to the curve generation algorithm, such as smoothing and outlier-removal, will result in a more precise average estimation.
- Investigating a way to represent technology non-adopters, including estimations from census data and/or time use surveys. Internet constrained areas would need identified as well, and a designated 'weighting' would fuse the different data sets.
- Performing comparative studies using objective building occupancy curves, generated from a properly implemented, active access-control policy. This would be expensive, time consuming, and require opt-in from building partners.
- Creation of a confidence scale for the crowd sourced estimations; which would require an objective baseline from the aforementioned comparative studies. Different classes of buildings would be delineated based on size, traffic, location, popularity, etc. Once complete, a structure's curve estimation confidence, or error estimation, can be gained by comparing with buildings in its class.

Twitter and Instagram were chosen for the services' consumer acceptance in the geographic area of San Jose, CA. Integration/weighting of other geospatially enabled social media products such as Foursquare, Facebook, Panaramio, etc., as well as data fusion with the Google PT curves, would result in a more robust estimation from social media.

6   Conclusions

This investigation has shown the use of open web services and enterprise systems to create occupancy curves for structures. Five additional buildings had curves generated, and their data is displayed. Comparisons with the Google PT functionality are shown and discussed for two buildings. While this social media aggregation method of estimating building occupancy is promising, further work is required to provide a reliable population metric.

References

1. About — Children's Discovery Museum of San Jose. https://www.cdm.org/about/. Accessed 3 Jul 2016
2. Aubrecht, C., Ungar, J., Freire, S.: Exploring the potential of volunteered geographic information for modeling spatio-temporal characteristics of urban population: a case study for Lisbon Metro using foursquare check-in data. In: International Conference Virtual City and Territory, Lisboa, vol. 7, pp. 57–60 (2011)

3. Batista e Silva, P., Martens, L.: Population estimation for the urban atlas polygons. Report no. EUR 26437 EN. European Commission, Joint Research Center, Ispra, Italy (2013). Print ISBN 978-92-79-35089-4
4. Bhaduri, B., et al.: LandScan USA: a high-resolution geospatial and temporal modeling approach for population distribution and dynamics. GeoJournal **69**(1–2), 103–117 (2007)
5. Center for the Performing Arts - San Jose Theaters. http://sanjosetheaters.org/theaters/center-for-performing-arts/. Accessed 3 July 2016
6. Convention Center — San Jose - Innovation Starts Here — Team San Jose 2015. http://www.sanjose.org/plan-a-meeting-event/venues/convention-center. Accessed 19 Sept 2015
7. Dong, B., Andrews, B.: Sensor-based occupancy behavioral pattern recognition for energy and comfort management in intelligent buildings. In: Proceedings of Building Simulation, pp. 1444–1451 (2009)
8. EPSG: 3857 - OpenStreetMap Wiki (2015). http://wiki.openstreetmap.org/wiki/EPSG:3857. Accessed 19 Sept 2015
9. Event Center Arena - Wikipedia, the free encyclopedia. https://en.wikipedia.org/wiki/Event_Center_Arena. Accessed 19 Sept 2015
10. Freire, S., Florczyk, A., Ferri, S.: Modeling day-and nighttime population exposure at high resolution: Application to volcanic risk assessment in Campi Flegrei. In: 12th International Conference on Information Systems for Crisis Response and Management (2015)
11. Freire, S., Florczyk, A., Pesaresi, M.: New multi-temporal global population GridsApplication to Volcanism. In: 13th International Conference on Information Systems for Crisis Response and Management (2016)
12. Fuchs, G., Andrienko, N., Andrienko, G., Bothe, S., Stange, H.: Tracing the German centennial flood in the stream of tweets: first lessons learned. In: Proceedings of the Second ACM SIGSPATIAL International Workshop on Crowdsourced and Volunteered Geographic Information, pp. 31–38 (2013)
13. GNIP Representative: Re: Twitter Data Discussion. Message to the author. E-mail (2015)
14. GNIP - The World's Largest, Most Trusted Provider of Social Data: The Source for Social Data. Accessed 17 Oct 2015
15. Google App Engine: Platform as a Service - App Engine Google Cloud Platform. https://cloud.google.com/appengine/docs. Accessed 19 Sept 2015
16. Grimes, J.G: Global Positioning System Standard Positioning Service Performance Standard. GPS Navster Department of Defense (2008)
17. Kuan, J.: Learning Highcharts 4. Packt Publishing Ltd. Accessed 19 Sept 2015
18. Kubanek, J., Nolte, E.-M., Taubenböck, H., Wenzel, F., Kappas, M.: Capacities of remote sensing for population estimation in urban areas. In: Bostenaru Dan, M., Armas, I., Goretti, A. (eds.) Earthquake Hazard Impact and Urban Planning. EH, pp. 45–66. Springer, Dordrecht (2014). doi:10.1007/978-94-007-7981-5_3
19. La Victoria Taqueria - 405 Photos - Mexican - Downtown - San Jose, CA - Reviews - Menu - Yelp. http://www.yelp.com/biz/la-victoria-taqueria-san-jose-2. Accessed 19 Sept 2015
20. Laituri, M., Kodrich, K.: On line disaster response community: people as sensors of high magnitude disasters using internet GIS. Sensors **8**(5), 3037–3055 (2008)
21. Martin, D., Cockings, S., Harfoot, A.: Development of a geographical framework for census workplace data. J. Roy. Stat. Soc.: Ser. A (Appl. Stat.) **176**(2), 585–602 (2013)
22. Martin, D., Cockings, S., Leung, S.: Developing a flexible framework for spatiotemporal population modeling. Ann. Assoc. Am. Geogr. **105**(4), 754–772 (2015)
23. Mennis, J., Hultgren, T.: Intelligent dasymetric mapping and its application to areal interpolation. Cartography Geogr. Inf. Sci. **33**(3), 179–194 (2006)
24. Namiot, D., Sneps-Sneppe, M.: Geofence and network proximity. In: Balandin, S., Andreev, S., Koucheryavy, Y. (eds.) NEW2AN/ruSMART -2013. LNCS, vol. 8121, pp. 117–127. Springer, Heidelberg (2013). doi:10.1007/978-3-642-40316-3_11

25. Oracle Technology Network for Java Developers — Oracle Technology Network — Oracle. http://www.oracle.com/technetwork/java/index.html. Accessed 19 Sept 2015

26. Popular times - Google My Business Help. https://support.google.com/business/answer/6263531?hl=en. Accessed 19 Sept 2015

27. Richardson, I., Thomson, M., Infield, D.: A high-resolution domestic building occupancy model for energy demand simulations. Energy and buildings **40**(8), 1560–1566 (2008)

28. Rose, A.N., Bright, E.A.: The LandScan Global Population Distribution Project: Current State of the Art and Prospective Innovation. Oak Ridge National Laboratory (ORNL) (2014)

29. San Jose, CA - Official Website - City Hall. http://www.sanjoseca.gov/Index.aspx?NID=233. Accessed 3 July 2016

30. San Jose Repertory Theatre - Wikipedia, the free encyclopedia. https://en.wikipedia.org/wiki/San_Jose_Repertory_Theatre. Accessed 3 July 2016

31. Sims, K.M., Weber, E.M., Bhaduri, B.L., Thakur, G.S., Resseguie, D.R.: Application of social media data to high-resolution mapping of a special event population. In: Griffith, D.A., Chun, Y., Dean, D.J. (eds.) Advances in Geocomputation. AGIS, pp. 67–74. Springer, Cham (2017). doi:10.1007/978-3-319-22786-3_7

32. Smith, A.: US Smartphone Use in 2015. Pew Research Center (2015)

33. South Hall — San Jose - Innovation Starts Here — Team San Jose. http://www.sanjose.org/plan-a-meeting-event/venues/south-hall. Accessed 19 Sept 2015

34. Student Union, Inc. of SJSU — San Jose State University. http://www.sjsu.edu/studentunion/. Accessed 3 July 2016

35. The Tech Museum of Innovation - Wikipedia, the free encyclopedia. https://en.wikipedia.org/wiki/The_Tech_Museum_of_Innovation. Accessed 19 Sept 2015

36. Toepke, S.L., Scott Starsman, R.: Population distribution estimation of an urban area using crowd sourced data for disaster response. In: 12th International Conference on Information Systems for Crisis Response and Management (2015)

37. Toepke, S.: Structure occupancy curve generation using geospatially enabled social media data. In: Proceedings of the 2nd International Conference on Geographical Information Systems Theory, Applications and Management, pp. 32–38 (2016)

Event Categorization and Key Prospect Identification from Storylines

Manu Shukla[1,2,3]([⊠]), Andrew Fong[1,2,3], Raimundo Dos Santos[1,2,3],
and Chang-Tien Lu[1,2,3]

[1] Omniscience Corporation, Palo Alto, CA, USA
{manu.shukla,andrew.fong}@omni.sc
[2] US Army Corps of Engineers ERDC GRL, Alexandria, VA, USA
raimundo.f.dossantos@erdc.dren.mil
[3] Computer Science Department, Virginia Tech, Falls Church, VA, USA
ctlu@vt.edu
http://www.omni.sc

Abstract. Event analysis and prospect identification in social media is challenging due to endless amount of information generated daily. While current research focuses on detecting events, there is no clear guidance on how those events should be processed such that they are meaningful to a human analyst. There are no clear ways to detect prospects from social media either. In this paper, we present DISTL, an event processing and prospect identifying platform. It accepts as input a set of storylines (a sequence of entities and their relationships) and processes them as follows: (1) uses different algorithms (LDA, SVM, information gain, rule sets) to identify themes from storylines; (2) identifies top locations and times in storylines and combines with themes to generate events that are meaningful in a specific scenario for categorizing storylines; and (3) extracts top prospects as people and organizations from data elements contained in storylines. The output comprises sets of events in different categories and storylines under them along with top prospects identified. DISTL uses in-memory distributed processing that scales to high data volumes and categorizes generated storylines in near real-time.

1 Introduction

In social media channels like Twitter, emerging events propagate at a much faster pace than in traditional news. Combining relevant facts together (while discarding unimportant ones) can be very challenging because the amount of available data is often much larger than the amount of processing power. This implies that many systems are unable to keep up with increasingly large volumes of data, which may cause important information to be missed. Event processing, therefore, is at a minimum dependent on two tasks: (1) collecting all the facts, entities, and their relationships; (2) grouping them by their themes of discussion, space, and timeframes. These two tasks should be performed in a distributed paradigm for maximum coverage. In the real world, not every piece of information

C. Grueau et al. (Eds.): GISTAM 2016, CCIS 741, pp. 62–88, 2017.
DOI: 10.1007/978-3-319-62618-5_5

can be thoroughly investigated in a timely manner. The goal, therefore, is to maximize the two previous tasks so that an event can be described with the most number of pertinent facts that yields the most complete picture. Figure 1 provides a visual representation of the idea. The figure shows seven tweets with a connection to the Boston area: t2, t3, and t7 are related to the Boston Marathon Bombings of April 2013, while t1 and t5 are about baseball, and t4 and t6 are about finance. First, these messages are certain to come hidden among millions of other tweets of different natures. Further, they relate to different topics, which indicates they should be presented separately. As seen in Fig. 1, all of the tweets are first transformed into simple storylines, and then grouped into three different themes ("Boston Marathon Bombings", "Wall Street News" and "Boston Red Sox"), which may be better suited to present to different audiences.

Social media is also a key place to identify emerging prospects for investment, partnerships and acquisitions [31]. For key prospect identification, finding top people and organizations from storylines model or performing unsupervised learning can be crucial for analyst. Figure 2 provides a visual representation of the idea. The figure shows three tweets (t1, t2, and t3) related to connected cars, self driving cars and industry activity. As Fig. 2 shows, all of the tweets are first transformed into simple storylines, and then four different prospects identified, three prospects represent large companies (Alibaba, BMW and Intel) and one (Mobileye) is a small company. The latter may be of particular interest to an analyst looking to invest in the connected car business, while other three may be better suited for analyst tracking large investments in the area.

The goal of this paper is to perform the above tasks using DISTL, Distributed In-memory Spatio-Temporal storyLine categorization platform (also shown in Fig. 1) [29], a system that ingests storylines derived from tweets, and allocates them to appropriate events along with finding top prospects for investments and acquisition in a given domain. The criteria used for the allocation process is that storylines have common themes, are located in nearby areas, and take place during close timeframes. DISTL uses as input the storylines generated by DISCRN [22], and is an in-memory spatio-temporal event processing platform that can scale to massive amounts of storylines using Big Data techniques. The platform helps analysts find faint, yet crucial events by separating storylines into groups, which allow analysts to sift through them in subsets under specific contexts.

A storyline is simply a time-ordered connection of facts that take place in a geographical area. In Fig. 1, for example, "police $\rightarrow$ block off $\rightarrow$ downtown Boston" represents a simple storyline related to a bigger event (the Boston Marathon Bombings). Storylines may be variable in length, and made as elastic as desired. In this paper, we do not show how these storylines are generated. Rather, we refer the reader to our previous work, DISCRN, which is a distributed platform specifically dedicated to generating storylines.

In order for a storyline "to be told", the user must first select a starting entity, such as a person or organization, from where the story can investigated. By checking the connections from that starting entity to other entities, one can then combine the facts together into a bigger event. For example, one may select a "person carrying a back pack" from one tweet to be the starting entity, and obtain other facts from other tweets, such as "entering subway", and "making a phone call", which would paint a more complete picture of a possible crime. DISCRN is a distributed system that mines storylines, as described above, at scale. It is effective in extracting storylines from both short unstructured tweet snippets and structured events such as in GDELT (Global Databases of Events, Language and Tone) [13]. DISCRN uses MapReduce [9] to generate all storylines from a specified starting entity from a large set of tweets. Since MapReduce is disk-based, it becomes less than ideal for highly-iterative event processing algorithms used in DISTL. For that reason, it is imperative to explore memory-based solutions explained later.

The key contributions of the platform are:

- **Design a Framework to Generate Spatio-temporal Events and Investment Prospects from Storylines.** Multiple algorithms (LDA, Information gain, classification) are applied to determine events that incorporate themes, location and time elements. The framework also identifies prospects for investment, partnership or acquisition for a domain.
- **Distribute In-memory Event Processing and Prospect Identification.** In-memory distributed architecture scales highly iterative event and prospect generation. This allows for processing large number of storylines efficiently.
- **Develop Rules based Technique to Categorize Storylines into Events.** Rules allow user fine-grained control on incorporating storylines into events. This provides user flexibility in focusing on theme, location or time.
- **Conduct Extensive Experiments to Validate Events Categorization and Top Prospects Identification from Storylines.** Framework and algorithms are validated with extensive experiments in multiple domains. The results effectively categorized storylines under meaningful events and found useful prospects for investigation.

The rest of the paper is organized as follows. Section 1 provides an overview of storylines and event creation with them. Section 2 describes the related works on event creation from social media data. Section 3 describes the techniques in DISTL in detail and Sect. 4 describes the architecture of components used to perform it. Section 5 presents experiments performed with datasets on different storyline subjects and themes on which meaningful and interesting events and top prospects were generated. Section 6 provides conclusions of the study.

2   Related Works

This section provides related works in event creation and distributed in-memory algorithms. Event creation in social media is a widely researched field. Event creation consists of identifying the event and characterizing it. Previous work primarily focuses on detecting events instead of categorizing elements under them.

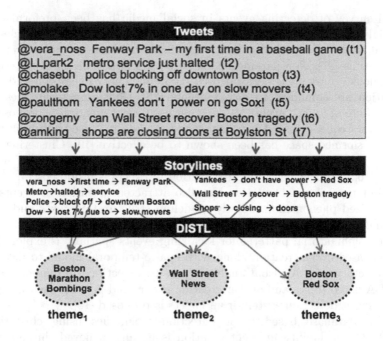

Fig. 1. Events used to categorize storylines.

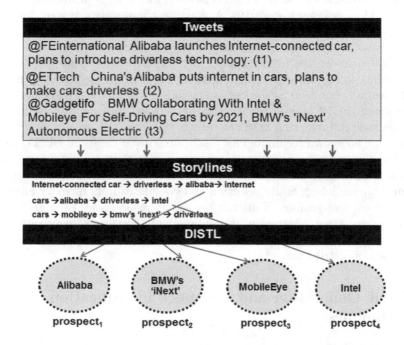

Fig. 2. Prospects from tweets used to build storylines of interest.

A useful survey of detecting events in social media has been preformed [11]. Inferential models are explored for detecting events from news articles [18]. Graphical models are used to detect events from multiple user posts over social media streams [27]. Non-parametric based scan statistic based event creation as anomalous subgraphs is performed [7]. Dynamic query expansion and anomaly creation are combined to detect events [25]. Clustering techniques along with signature of known events based supervised event creation is proposed [1]. Wavelets on frequency based raw signals and modularity based graph partitioning on transformed space has been shown to be effective [10]. Clustering based sub-event identification is investigated [17]. Tweets as replacement of news is explored [16]. Segment based event detection in tweets is proposed [14].

Spatio-temporal event creation has also attracted attention as events tend to be localized phenomenon. Jointly modeling structural contexts and spatio-temporal burstiness is used in event forecasting [26]. Burstiness and spatio-temporal combinatorial patterns for localizing events spatially is explored [12]. Classifier based event creation along with spatio-temporal model to determine event trajectory [21]. A visual analytics way to detect spatio-temporal events using LDA based topics and outlier creation is explored [6]. A sequential spatio-temporal event detection system from tweets is proposed [15].

NoSQL database based event detection techniques using clustering is explored [23]. Scalability in event creation is usually achieved through transforming the problem to efficient domains. Scalability in event creation from social media streams with event based clustering by reducing problem to a record linkage problem is investigated [19]. A scalable non-negative matrix factorization based technique to detect events in social media is presented [20]. However, in case of storytelling, events have to be generated such that all storylines are attributed under the event making it imperative that none are dropped. That requires scaling through distribution rather than problem transformation. Machine learning has been used widely to manage investments. Financial Genetic Programming (FGP) is used for investment decision making [28]. Neural nets have been used in predicting takeover targets [30].

There are no known techniques for distributed event creation. DISTL applies highly iterative techniques to event theme generation that can not be scaled efficiently with disk based distribution such as MapReduce. Use of Apache Spark to perform topic modeling, entity selection and classification in memory allows for much more efficient scaling. It distributes the entire sequence of steps starting from composite event generation and subsequent storyline categorization into those events in-memory. This allows to scale the process completely and maximize impact of distribution.

3   Event Generation and Prospect Identification Techniques

In this section the techniques used to generate events from storylines and categorize storylines under those events are described. Subsection 3.1 provides brief overview of distribution techniques in Spark. Subsection 3.2 presents theme generation technique followed by Subsect. 3.3 that explains how events are generated from themes and storylines assigned to the events. Subsection 3.4 describes the top prospect identification techniques in DISTL.

3.1 In-Memory Distribution in Spark

Apache Spark is an in-memory distribution framework that allows computations to be distributed over a large number of nodes in a cluster [24]. The programming constructs available in Spark represent transformation of data on disk into RDDs (Resilient Distributed Datasets), which reside in-memory. Operations applied on the RDDs to generate values that can be returned to the application. RDDs provide fault tolerance in case one or more nodes of the cluster fail. The algorithms typically useful for Spark are the ML (Machine Learning) and statistical functions that are highly iterative in nature. Performing highly distributed operation in any other distributed construct such as MapReduce is computationally expensive due to data written to disk in each iteration. Spark allows for highly efficient iterative operations as they are all performed in memory.

The main operations provided by Spark that allows it process large volumes of data in parallel can be broadly categorized into *actions* and *transforms* [4]. The *transform* operations commonly used include *map, filter, flatMap, join, reduceByKey* and *sort*. The *action* operations commonly used are *reduce, collect* and *count*. The *map* operation returns a distributed dataset by applying a user specified function to each element of a distributed dataset. A *flatMap* does the same except one input term can be mapped to multiple output items. *reduceByKey* operation aggregates the values of each key in key-value pairs <K,V> according to provided reduce function. *filter* returns datasets from source for which given function evaluates true. Spark also allows to keep a read-only value of variables in cache on each node instead of shipping them with each task through broadcast variables.

3.2 Theme Creation

Several major theme recognition techniques are made available to the analyst. The event creation technique uses top-weighted keywords as themes. A dictionary-based method assigns storylines to event buckets. Rule based storylines categorization is performed. We can generate events based on theme, location and time. The dictionary is generated for themes by analyzing the terms of the storylines and discovering key ones. The recognized themes are then used to categorize the storylines.

The key aspect of event generation is identifying the entities that are closest to significant events. The sequence of steps in events generation and assigning storylines to events is shown in Fig. 3. The flow consists of 3 main steps; process storylines, build themes and create events, and score and categorize storylines. The first step processes storylines and identifies spatial and temporal entities in them. Here supervised and unsupervised techniques are used in the identification of the most critical entities, which are used in the subsequent step. Theme, spatial and temporal entities are combined to generate events. The storylines are then categorized under the events in the last step of flow. The three algorithms used in second step of flow are as follows.

- Topic Modeling based event creation: Topic models are generated using Latent Dirichlet Allocation (LDA). It uses a list of storylines as input data, where each storyline represents a document. This technique outputs a number of topics specified by the user, where each topic is comprised as a set of keywords.
- Feature Selection through Information Gain based event creation: This technique extracts top n keywords by information gain from storylines. Each storyline is treated as a document. Each of the highest n information gain value keywords is treated as belonging to the subject for which labeled data was generated.
- Classifier based event creation: This technique uses a classifier trained with user generated set of storylines for a particular subject. This model is then used to classify storylines into ones belonging to that subject or not. An example would be if analyst wants to separate all storylines that are related to earnings of a company from ones that are not. A classifier based technique works best in case of known subjects being analyzed in storylines. Events under which storylines are categorized are generated using most frequent theme, location and time entities in positively labeled training data. Storylines are also scored based on the classifier score or the score of their entities for topic or feature selection. These scores are then used to filter top storylines by applying a threshold. Data elements for entities of top storyline are then identified and used to create top people and organizations as prospects.

Topic modeling falls under unsupervised learning while other two (information gain and classifier) are supervised. They require training data in order to generate the themes. All these techniques are highly iterative and under large datasets computationally expensive especially in terms of building model. Algorithm 1 shows the application of 3 techniques to categorize events. Step 1 performs the extraction of entities from storylines and generating RDDs of storylines from JSON output produced by DISCRN. One RDD is created for training data and one from scoring data. A combined index of entities is generated. Step 2 then generates RDDs of theme entities and other entities identified as location and time as PairRDDs. PairRDDs are <Key, Value> pair representation in RDDs. It then performs LDA based topic modeling, feature selection based on information gain or SVM model generation and most frequent entities from positively labeled training data to extract themes from entities. All operations are implemented such that they are performed in memory.

3.3   Event Generation and Storylines Assignment

Events are generated by combining themes with the spatial and temporal entities identified in storylines. Algorithm 2 shows how generating events based on

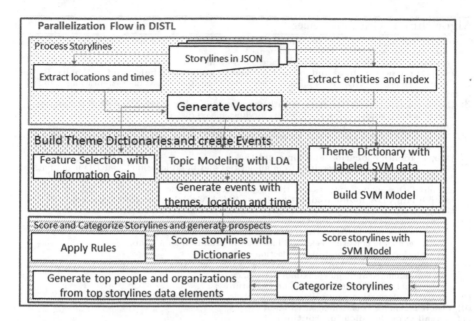

Fig. 3. DISTL system flow.

themes, location and time entities is performed in-memory with the location, time and theme entities extracted from entities RDD and then combined together to create events. The task of finding the combinations of location, time and entities based on one or more subject depends on the technique used for subject creation or labeled data based entity extraction. These are crucial to identifying events and associating entities with events. Step 2 categorizes storylines into events. This approach tests keywords in a storyline against the spatial, temporal and theme entities and assigns it to the theme based events using rules specified by user.

The rules provided by user to categorize storylines into events with theme, location and time elements and their application on storylines is explained below.

Categorization Rules Format. The rules are of following format:
theme ([*and*|*or*] location [*and*|*or*] time)
Hence the rules can take any of the following forms:

1. theme *or* (location *and* time)
2. theme *or* location
3. location
4. time

The rules specify which entities in a storyline need to match with rule entity of particular type in order to associate a storyline to the event. Hence Rule 1 specifies that only if storyline entities match either theme or location and time

---

Algorithm 1. Generate themes.

Input: $\{storyline_i\},\{labeled_j\}$ {unlabeled storylines and labeled storylines for supervised learning}
Output: $\{event_k, storyline_i\}$ under each $event_k$}
    {each event definition and storylines under each event}
 1: {**step 1: parse storylines and extract entities to generate labeled and unlabeled RDDs**}
 2: Create trainingDataRDD from labeled storylines file on distributed file system using map transform
 3: Create entityIndexRDD as index of entities to integers using flatMap and filter transforms
 4: Create testingDataRDD from unlabeled storylines file on distributed file system using map transform
 5: Create labeledVectorsRDD and unlabeledVectorsRDD with vectors for storylines using zipwithindex and distinct transforms
 6: {**step2: Identify location and time entities** }
 7: Extract location and time entities from all entities and build locationTimeRDD using flatMapToPair transform
 8: as PairRDD as map of entities and their type location or time
 9: {**step3: run LDA model and get topics or SVM Modeling or Info Gain Feature Selection**}
10: **if** technique is topic modeling **then**
11:    ldaModel = new LDA().setK(noOfTopics).run(storylineEntityVectorsRDD);
12: **else if** technique is feature selection based on information gain **then**
13:    Create FeatureSelection featureSelection object
14:    Perform MRMR feature selection by featureSelection.minimumRedundancyMaximumRelevancy (storylineEntityVectorsRDD, numberOfFeatures);
15:    extract information gain values by featureSelection.relevancyRedundacyValues();
16: **else if** technique is classifier based **then**
17:    {**call svm classification routine**}
18:    Build SVMModel by invoking SVMWithSGD.train(labelPointsRDD, numIterations)
19:    score testingDataRDD with model
20:    Extract themes as PairRDD¡String,Integer¿ from positively scoring data
21: **end if**

Algorithm 2. Generate Events.

Input: $\{storyline_i\},\{labeled_j\}$ {storylines and labeled storylines for supervised learning}
Output: $\{event_k\}$
    {each event definition and storylines under each event}
 1: {**step 1: Use themes and dictionaries generated in previous algorithm**}
 2: Get locTimeRDD from previous step
 3: Get labeledVectorsRDD from previous step
 4: {**step 2: Use output from applied algorithm from previous step**}
 5: **if** technique used is topic modeling **then**
 6:    {**Applying top LDA weighted themes, locations and times**}
 7:    **for all** topic $\in$ Topics **do**
 8:      Extract top location, time and theme term along with their weights
 9:      Combine top weighted theme, time and location entity into event
10:    **end for**
11:    Get k events where k were number of topics extracted
12: **else if** technique is feature selection based on information gain **then**
13:    {**Generate events with top info gain entities**}
14:    Generate event as combination of top information gain theme, location and time
15: **else if** technique is classifier based **then**
16:    {**Generate events with top positively labeled storylines location, time and theme entities by frequency**}
17:    Calculate frequency of entities in positively labeled documents
18:    Combine top location, time and theme entities into events
19: **end if**

---

then categorize the storyline to the event. Rule 2 can categorize a storyline to the event if any of its entities matches either theme or location of the event while Rule 4 associates any storyline whose entities match the temporal entity of the event.

Rules Application. As each of these rules are applied to a storyline for each event, if any rule is satisfied for a storyline against an event, the storyline is categorized under that event. Algorithm 3 categorizes storylines under events applying the rules. Rules are broadcast to all the nodes and the storylines RDD then has each storyline in it run through the rules and associated with an event if any rule matches the storyline to the event. As soon as a storyline is associated with an event the rules application ends. Based on number of entities in a storyline that match rule's theme, location or time, a weight is assigned to storylines. For classifier events the weights are normalized with the storylines classifier score.

3.4 Prospect Generation from Storylines

Prospects are generated by extracting key people and organization from data elements used to build storylines. Algorithm 4 shows how generating prospects from tweets of storylines entities is performed in-memory with the tweets from entities RDD. The task of finding top scoring storylines utilizes the weights from topic modeling and features selection and the storylines scores against SVM classifiers as shown in step 2. These are then utilized to generate tweets from top storyline entities and extract top people and organizations as prospects in step 3.

4 System Architecture

This section describes the overall architecture of DISTL. The Subsect. 4.1 describes theme and dictionary creation component while Subsect. 4.2 describes the component that categorizes storylines into events. Prospect identification component is detailed in Subsect. 4.3.

4.1 Theme and Dictionary Creation

The system architecture of the platform to generate events in storylines is shown in Fig. 4. Due to large number of storylines generated from tweets collected on topics, the amount of data to be processed to generate events on the entities can be large. Event creation is performed as an extension to the DISCRN platform. In-memory distribution is essential to computing topics and perform feature selection based on information gain as these techniques tend to be highly iterative and do not scale well on disk based distribution paradigms such as MapReduce as disk I/O will be highly detrimental to performance. The modules described in this subsection generate themes from storylines, identify location and time entities and combine them to create composite events.

- Process Storylines: This job in Spark reads the storylines in parallel and extracts entities from them. Vectors are built with indices of entities in storylines.

Algorithm 3. Categorize storylines under Events.

Input: $\{storyline_i\}$,$\{event_j\}$,svmModel {storylines and labeled storylines for supervised learning}
Output: $\{event_k,storyline_i$ under each $event_k\}$
　　{each event definition and storylines under each event}
1: {**step 1: parse rules**}
2: Broadcast rules to all worker nodes
3: Read rules in broadcast var
4: {**step 2: Apply rules to generate events depending on algorithm previously applied**}
5: **if** technique used is topic modeling **then**
6: 　　{**Categorize storylines under topic events**}
7: 　　**for all** topic $\in$ Topics **do**
8: 　　　　PairRDD<Integer, Storyline>topicToStoryLinesRDD using mapToPair transform by apply-
　　　　ing rules and dictionaries by topic to storylines
9: 　　**end for**
10: **else if** technique used is feature selection **then**
11: 　　{**Categorize storylines under feature selection events**}
12: 　　Build RDD fsStoryLinesRDD using map transform by applying rules and events to storylines
13: **else if** technique is svm **then**
14: 　　{**Categorize storylines under svm events**}
15: 　　Build RDD classifierStoryLinesRDD using map and filter transforms by applying rules and
　　　scored storylineVectorRDD against model
16: 　　**if** score $\geq$ threshold and match rules **then**
17: 　　　　assign storyline to event
18: 　　**end if**
19: **end if**

Algorithm 4. Generate Prospects.

Input: $\{storyline_i\}$,$\{labeled_j\}$ {storylines and labeled storylines for supervised learning}
Output: $\{prospect_k\}$
　　{top scoring storylines against model in supervised learning or in unsupervised learning and the
　　data elements they were built from}
1: {**step 1: Use dictionaries generated in previous algorithm**}
2: Get labeledVectorsRDD from previous step
3: {**step 2: Use output from applied algorithm from previous step to score storylines**}
4: **if** technique used is topic modeling **then**
5: 　　{**match entity in each topic with storylines entity**}
6: 　　**for all** topic $\in$ Topics **do**
7: 　　　　{**Generate top scoring storylines by topic based on topic weights of entities**}
8: 　　　　Score storylines entities against topic entity and sum weights of entities in topic
9: 　　　　Build list of top K storylines by topic
10: 　　**end for**
11: **else if** technique is feature selection based on information gain **then**
12: 　　{**Generate top scoring storylines with top info gain entities**}
13: 　　score each storyline as a combination of its entities info gain weight
14: **else if** technique is classifier based **then**
15: 　　{**Generate top scoring positively labeled storylines** }
16: 　　Get entities of top scoring storylines against classifier
17: **end if**
18: {**step 3: Generate list of tweets from entities of top storylines and get most frequent
　　people and organization from tweets**}
19: Lookup tweets associated with the top M storyline entities
20: PairRDD <String, Integer>as count of people and organizations from tweets of top storyline
　　entities

– Determine Spatial and Temporal terms: This module determines the spatial
and temporal terms using the GATE API [8]. Each storyline is broken down
into entities in parallel and in each process GATE APIs are initiated and
used to label entities in the storyline document. The entities identified in
processing step are used to create an index of entity strings to integers that
is then used on storyline vectors in subsequent step.

– Build Themes and theme dictionary terms: The vectors built in processing step for each storyline are passed to one of the three theme generation routines.

1. Topic Modeling based: When theme building process specified is topic modeling the vectors are passed to the MLLib LDA based topic modeling technique [2]. This technique returns the entities for the topics and their corresponding weights for the topic. These are then saved as dictionary for the theme.
2. Entity Selection based: If the specified theme building process is entity selection based on information gain, the vectors are passed into the information gain based entity selection routine based on Maximum Relevancy Minimum Redundancy [5]. This technique performs information gain in parallel to generate a list of top k entities for the labeled training set. This list is saved as the dictionary for the theme event on which the labels in training data are based.
3. Classification based: For chosen theme building process classification, the labeled data for storylines is used to build an SVM model using the MLLib SVM Spark routine [3] that build the model in parallel. This model is then used to score the storylines and top k positively labeled storylines entities are chosen and added to the dictionary.

4.2 Events Creation and Storyline Categorization

These modules assign storylines to generate events in a scalable way. Storylines event assignment module scores each storyline and determines which event they will be assigned to based on user specified rules.

– Generate Events: This module generates events as a combination of themes and spatiotemporal entities.
1. Topic Modeling: In case of topic modeling an event is generated for each topic with the top theme entity of the topic, top location entity and top time entity by weight combined to generate the spatio-temporal event.
2. Feature Selection: In case of feature selection top weighing theme, location and time entity with highest information gain value are combined to generate the event that corresponds to class of labeled data.
3. Classifier: In case of classifier, the most frequent theme, location and time in positively labeled training data is combined to generate event.
– Test storylines: This job loads theme, location and time dictionaries in cache and is used to test storylines in parallel to identify storylines that fall within the event. Rules provided by end user are broadcast to all nodes and processed and information in the rules is used to determine how to assign storyline to event.
– Categorize storylines into events: This job categorizes and lists in parallel events and the storylines within them. Events are a combination of theme, location and time in format *theme:location:time*.

4.3   Prospect Identification

These modules identifies top storylines and computes the tweets associated with the entities in those storylines. It then applies named entity recognition using GATE on the text of the tweets to generate the top people and organizations. Storylines are scored in each of the techniques using the weight of entities in topic, the score assigned to storyline by classifier and the weight associated with entities in MRMR feature selection.

– Generate Top Storylines: This module generates top storylines from SVM scores in case of classification, entity weight in topic in topic modeling and feature weight in feature selection.
  1. Topic Modeling: In case of topic modeling a storyline is associated with a score that is the sum of the entity weights for each topic.
  2. Feature Selection: In case of feature selection storyline is given a score which is the sum of the score of the entities of storyline.
  3. Classifier: In case of classifier, the storyline is scored against SVM model and is assigned a score.
– Collect tweets associated with storyline entities: This job generates a list of tweets associated with each entity of a storyline.
– Extract named entities from tweets as people and organizations: This job generates a list of people and organizations that are in the list of tweets associated with top scoring storylines by technique and lists them by frequency.

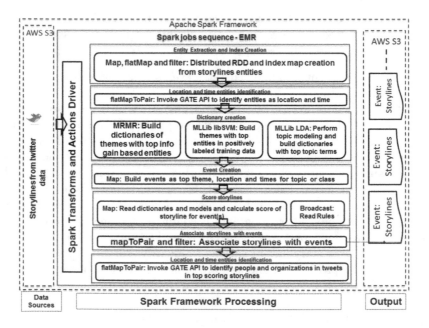

Fig. 4. DISTL architecture.

5   Use Cases

This section describes the use cases for event categorization and finding prospects in various distinct domains. Section 5.1 describes the event categorization use cases while Sect. 5.2 illustrates finding top prospects through DISTL. Section 5.3 presents performance evaluation results.

5.1   Event Categorization Use Cases

Extensive experiments were performed to validate event creation and storylines categorization as scalable and useful to analysts. Two sets of storylines built on Twitter data extracted using keywords filtering using Twitter streaming API in June 2015 were used. This section describes events generated for commodity oil and events on currency Euro.

The topic modeling technique uses LDA and produces topics with top keywords and weights added to topic's dictionary. After experiment with several topic numbers, the subject expert decided on 3 as optimal number of topics from which to generate events. That number was within the constraints of not subdividing an event into subevents and yet generate meaningful results. Feature Selection based on Information gain generates a dictionary of themes with highest information gain. The features selection routine is given a set of labeled observations based on a specified class. The training data set is generated by a subject matter expert. For oil dataset a training set was built for storylines related to oil prices and for euro dataset the training data was for Greek exit from Euro. The classifier based themes dictionary creation uses labeled data provided by subject matter expert. It uses the training set to build a classifier. The most frequent features are used as dictionary and top theme, location and time terms are used to create the event. Storylines with score higher than a threshold as classified by model and entities satisfying the categorization rules are associated with the event. Storylines in each event are assigned by testing the entities against event's location, time and theme entities. The rule used to categorize storylines into events in our experiments is "location *or* time *or* theme".

Oil Events. This subject is regarding tweets related to the commodity oil. The filtering keywords for tweets extracted are 'oil, wti, brent, crude, frack, fracking, oil price, petroleum, petro, west texas intermediate, oil inventory, oil production, oil refinery, #crudeoil and nymex'. The entry points for storylines are 'oil' and 'petroleum'.

Applying topic modeling produces the topics shown in Table 1. The top entities for each topic are similar, two 'amp' and 'petroleum' being top entities in each of the 3 topics and the third ranked entities being number 24, gas and canvas indicating one of the topics is more related to painting oil. Applying information gain based feature selection generates the top terms shown in Table 1. These weights more accurately reflect oil related storylines as expected of a supervised technique as the top information gain weight terms are not only related to oil but

Table 1. Oil topic modeling, info gain based feature selection and classifier labeled data based top entity weights.

Topic1	Weight	Topic2	Weight	Topic3	Weight
amp	0.033	amp	0.32	amp	0.0334
petroleum	0.023	petroleum	0.0266	petroleum	0.023
24	0.0150	gas	0.0129	canvas	0.011
Feature	Info gain	Feature	Info gain	Feature	Info gain
US $61.53	0.0335	weather fueling oil price discovery	0.0267	global crude oil price	0.0294
Singapore	0.006	oil prices	0.0288	17-Jun	0.0091
Feature	Weight	Feature	Weight	Feature	Weight
petroleum	333	US $61.53	58	oil price	77
London	44	reuters	61	17-Jun	15

also to the price of oil which was the basis of labeled data. Top features by classifier are in Table 1. These features accurately reflect the key entities in training dataset specifically the ones most frequent in positively labeled elements.

Top storylines based on events from top feature selection entities are in Table 2. Top storylines by weight for each topic from topic modeling as generated by application of rules are in Table 2. Top storylines based on events from classifier based top themes, locations and time entities are in Table 2. These categorizations clearly show that storylines in same set get categorized differently under different events simply due to application of different entity ranking techniques.

The analysis of oil brought mixed results, which is potentially explained by the broad range of topics covered by the term oil. For example, oil is associated with petroleum, body oil, oil paintings and other categories unrelated to the area of focus. Thus, the topics and associated storylines identified by topic modeling shown in Tables 1 and 2 hold limited relevance and add little to the understanding of oil prices. The results from topic modeling were also clouded by the entity 'amp' which actually refers to the character '&' and was erroneously picked up as an entity. However, the information gain and classification models did reveal interesting topics and storylines. First, the international crude oil price of Indian Basket as computed by the Ministry of Petroleum and Natural Gas of the Republic of India was $61.53 per barrel on June 16, 2015, an informative metric given India is one of the largest importers of petroleum in the world and the Indian government heavily subsidizes those imports. Second, the entity 'weather fueling oil price discovery' alluded to the foul weather moving through Texas at that time which was expected to impact oil production and thus prices.

Euro Events. This subject involves tweets regarding the change in value and status of the currency euro. Filtering keywords 'Euro, Euro exchange rate, euro

Table 2. Oil top storylines by LDA, Info Gain and classifier weight.

Topic event	Storyline	LDA weight
amp:north america:1988	oil:london #jobs:amp:sales manager	0.0422
amp:north america:1988	oil:amp:deals:funnel	0.0400
petroleum:london:17-Jun	oil:london #jobs:amp:sales manager	0.0422
petroleum:london:17-Jun	amp:seed oil unrefined 100:deals	0.0401
gas:greece:today	oil:amp:engine oil:gas	0.0480
gas:greece:today	oil:grease:paper 10 sets face care:amp	0.0470
Topic event	Storyline	Info gain weight
US $61.53:singapore:17-Jun	oil:petroleum:us $61.53:global crude oil price	0.0629
US$61.53:singapore:17-Jun	oil:petroleum:us $61.53:weather fueling oil price discovery	0.0602
US $61.53:singapore:17-Jun	oil:reuters:oil prices:production	0.0513
Topic event	Storyline	Classifier weight
petroleum:london:17-Jun	oil:long:iran:petroleum	1.2209
petroleum:london:17-Jun	oil:petroleum:oil price:our 2015 global #oil	1.1375
petroleum:london:17-Jun	oil:brent oil forecast. current price:petroleum:global crude oil price	1.1114

rates, euro-dollar, dollar euro, euro crisis, euro conversion, euro rate, eur and eur usd' are specified for tweet collection. The entry point for storylines is 'euro'.

On euro related storylines we applied event generation techniques. Applying topic modeling generates the topics shown in Table 3. Three topics were provided to the LDA method. The top entities were similar for the three topics with the difference being in third highest weighted entity indicating topics being related to emergency summit over Greek crisis. The top features by classifier are shown in Table 3. These features are more accurately related to the Greek exit due to the application of training data on the subject provided. Applying information gain based feature selection produces the top terms shown in Table 3. These entities are also highly relevant due to use of training data. Top storylines by weight for each topic as generated by application of rules are in Table 4. Top storylines based on events from top feature selection entities are given in Table 4. Top storylines based on events from classifier based top themes, locations and time entities are provided in Table 4. These storylines clearly show the preponderance of storylines on Greek exit crisis from the Euro at the time and the Federal Open Market Committee meeting on June 18, 2015.

On the analysis of the Euro dataset, topic modeling, information gain, and classification all highlighted the crisis occurring in Greece's economy and the potential of a Greek Exit from the Euro. Topic modeling even highlighted the emergency summit taking place in Luxembourg to discuss the situation. In this case, the information gain based feature selection analysis generated the most noise as the highest weighted features included indiscernible numbers and entities related to sports even though two of the features were 'greek exit' and 'syriza

Table 3. Euro top entities by info gain feature selection, topic modeling and classifier weights.

Topic1	Weight	Topic2	Weight	Topic3	Weight
The euro	0.0136	the euro	0.0137	the euro	0.0139
eur	0.012	eur	0.0123	eur	0.0121
Luxemborg	0.0055	emergency summit	0.0064	2015	0.008
Feature	Info gain	Feature	Info gain	Feature	Info gain
Greek exit	0.0621	0.049	0.0302	0.03	0.0461
18 june #football #soccer #sports	0.0282	0.08	0.0473	syriza hardliners back	0.0228
Feature	Weight	Feature	Weight	Feature	Weight
Greek exit	194	yesterday's fomc meeting	91	72.43	108
2015	24	1199.9	97	greece	83

hard-liners back'. The number of storylines an analyst has to review is greatly reduced for events, for SVM the number of storylines is reduced to 933 from over 300000 when threshold of 1.0 is set for the SVM scores.

5.2   Identify Key Prospects Use Cases

DISTL provides users with the ability to track investment and target prospects by identifying emerging people and organizations in open data. Further, DISTL narrows down the number of prospects down to the ones identified in the storylines with highest scores based on the models built to identify them. Two sets of storylines built on Twitter data extracted using keywords filtering using Twitter streaming API in July 2016 were used. Section 5.2 provides details of identifying top prospects within Autonomous Car domain and Sect. 5.2 provides details on top prospects for Augmented Reality/Virtual Reality (AR/VR) domain.

Autonomous Car Investments. This use case identifies top people and organizations with the highest impact in domains of connected car and autonomous driving. Filtering keywords 'Autonomous vehicle, Mobile Car Sharing, Telematics, driverless vehicle, ..., driver-assistance systems' are specified for tweet collection. The entry points for storylines are 'connected car, driverless car'.

On connected car related storylines we applied the three supervised and unsupervised techniques we had earlier used in event generation. Applying topic modeling generates the topics and entity weights as shown in Table 5 and corresponding storyline weights shown in Table 6. Three topics were provided to the LDA method. The top storylines by classifier weight are shown in Table 6. Applying information gain based feature selection produces the top entities in Table 5

Table 4. Euro top storylines by event and topic modeling, info gain and classifier weights.

Topic event	Storyline	LDA weight
the euro:2015:luxemborg	euro:zone ecofin meetings:the euro:eur	0.03106
the euro:2015:luxemborg	euro:dibebani yunani:eur:the euro	0.02900
eur:19-Jun:greece	euro:zone ecofin meetings:the euro:eur	0.0309
eur:19-Jun:greece	euro:dibebani yunani:eur:the euro	0.0288
amp:this day:edinburg	euro:zone ecofin meetings:the euro:1.7:0	0.0207
amp:this day:edinburg	euro:lows:6:eur	0.0215
Topic event	Storyline	Info gain weight
greek exit:greece:18 June	euro:2:0.08:#dollar	0.0068
greek exit:greece:18 June	euro:gold:0.13:0.08	0.029
greek exit:greece:18 June	oil:gas temp:marks sattin:#cash #applications accountant	0.29
Topic event	Storyline	Classifier weight
greek exit:greece:2015	euro:yesterday's fomc meeting:greek exit:support	1.3681
greek exit:greece:2015	euro:yesterday's fomc meeting:greek exit:greece #euro	1.3410
greek exit:greece:2015	euro:central bank:greeks themselves:greece	1.2701

Table 5. Connected car top entities by info gain feature selection, topic modeling and classifier frequency weights.

Topic1	Weight	Topic2	Weight	Topic3	Weight
connected car	0.1659	connected car	0.1626	connected car	0.1586
driverless car	0.0883	driverless car	0.0863	driverless car	0.0900
navigation	0.0597	Chicago	0.0542	navigation	0.0628
Feature	Info gain	Feature	Info gain	Feature	Info gain
1	0.0827	car	0.0327	venturecapital	0.0106
wingz	0.005	chicago	0.0071	0	0.0120
Feature	Frequency	Feature	Frequency	Feature	Frequency
China	194	alibaba	91	volkswagen	108
internet	180	reuters	97	safety	83

and top storylines shown in Table 6. These entities are also highly relevant due to use of training data.

Top storylines by weight for each topic as generated by application of rules are in Table 6. Top named entities in data elements for top storylines are shown in Table 7.

Table 6. Connected car top storylines by topic modeling, info gain and classifier weights.

Storyline	LDA weight	Topic
connected car:automotive:car platform:technews	0.2017	topic 1
connected car:lg:technology:the road	0.2010	topic 1
connected car:china:economy:new post	0.2021	topic 2
connected car:alibaba's internet:car:new post	0.2020	topic 2
connected car:jobs:northbrook:systems design	0.2022	topic 3
connected car:alibaba's internet:car:the world	0.2019	topic 3
Storyline	Info gain weight	
connected car:car:nokia growth partners:unquotenews	0.1162	
connected car:car:jaguar land rover's new tech subsidiary inmotion:smartcommuter	0.1051	
connected car:china:navigation:the navyo smart glove	0.0870	
Storyline	Classifier weight	
connected car:alibaba's internet:car:connected	0.1167	
connected car:automotive:programs:mapping	0.1154	
connected car:automotive:the google self:mapping	0.1145	

Results of the connected car analyses proved useful in terms of understanding the key connected car events of the day as well as identifying the individuals and organizations related to those events. Each algorithm produced different, yet complementary results that provided the analyst with a broad picture of the different themes emanating from the day's Twitter data. For instance, topic modeling highlighted LG's new partnership with Volkswagen to develop technology to allow drivers to monitor and control devices in the home from their car as well as improve in-vehicle entertainment capabilities. Topic modeling also highlighted the newly developed car with the YunOS operating system jointly developed by China's Alibaba's and SAIC Motors. The information gain algorithm produced different results that emphasized Jaguar Land Rover's new venture in the area car sharing through their wholly owned subsidiary InMotion and Nokia Growth Partner's new internet-of-things investment fund, which invests in connected car companies. The classification results partially overlapped with the topic modeling results, as both identified Alibaba's activities. And classification added additional value by also highlighting updates to Google's self-driving car initiative, which was not captured in top storylines from topic modeling. Ultimately, the list of storylines identified by the three algorithms was distilled to a list of persons and organizations found within the data underlying those storylines (Table 4). While some organizations were mischaracterized as persons, including ZF Friedrichshafen (a German auto-parts manufacturer) and BABA (the ticker symbol for Alibaba), the algorithm did an adequate job of identifying

Table 7. Connected car top named entities by info gain feature selection, topic modeling and classifier weights.

Organization	Topic frequency	Person	Topic frequency
LG	40	ZF Friedrichshafen	1
VW	9	Lee Coleman	1
Ford	2	ena	1
Organization	Info gain frequency	Person	Info gain frequency
LG	68	aren	3
Volkswagen	42	ena	1
Reuters	20	BABA	1
Organization	Classifier frequency	Person	Classifier frequency
LG	40	Ethan Lou	1
Reuters	20	Lee Colman	1
Tesla	4	aren	1
Google	4	Jake Spring	1

relevant and interesting entities. Tesla and Google's efforts in connected car are well known, but LG's initiatives were certainly less publicized. The individuals identified included connected car expert Lee Colman, Head of Connected Car at SBD Automotive (a UK based consultancy). Mr. Colman was also a frequent speaker at various conferences including CarConExpo Dusseldorf 2016 and Telematics Berlin 2016 and is, perhaps, an expert worth following. Other individuals included Ethan Lou and Jake Spring, both Reuters reporters who cover the car industry.

Augmented Reality/Virtual Reality Prospects. This use case identifies top people and organizations that are having the highest impact in nascent augmented and virtual reality domains. Filtering keywords 'virtual reality, augmented reality, ..., vr bandwidth, ar bandwidth, vr optimization, ar optimization, amazon vr, amazon ar' are specified for tweet collection. The entry points for storylines are 'vr, ar, virtual reality'.

On augmented reality related storylines we applied the three supervised and unsupervised techniques we had earlier used in event generation. Applying topic modeling generates the topics shown in Table 8. Three topics were provided to the LDA method. The top entities were similar for the three topics with the difference being in third highest weighted entity. The top features by classifier are shown in Table 8. Applying information gain based feature selection produces the top terms shown in Table 8. These entities are also highly relevant due to use of training data. Top storylines by weight for each topic as generated by application of rules are in Table 9. Top storylines based on events from top feature selection entities are given in Table 9. Top storylines based on events from classifier based top themes, locations and time entities are provided in

Table 8. AR top entities by info gain feature selection, topic modeling and classifier weights.

Topic1	Weight	Topic2	Weight	topic3	Weight
Virtual reality	0.0901	Virtual reality	0.1307	vr	0.1281
The future	0.0264	The future	0.0901	Virtual reality	0.0912
The magic leap	0.0260	Startup	0.0260	Startup	0.0297
Feature	Info gain	Feature	Info gain	Feature	Info gain
Virtual reality	0.0215	0	0.1083	1	0.0549
2016	0.005	vr	0.0207	Our blog	0.0110
Feature	Frequency	Feature	Frequency	Feature	Frequency
Games	58	Oculus	61	News	18
Reality	175	Headset	99	Pokemon	24

Table 9. AR top storylines by event and topic modeling, info gain and classifier weights.

Storyline	LDA weight	Topic
vr:cool:front:i recalibrate the oculus sensor	0.2031	topic 1
vr:cool:la france:metheniyacine	0.2023	topic 1
vr:google:reality #technology:tango:vr	0.2044	topic 2
vr:htc vive:people:robbers	0.2038	topic 2
vr:amazing:oculus:verder gaan	0.2021	topic 3
virtual reality:samsung:tech:three companies	0.2020	topic 3
Storyline	Info gain weight	
vr:eyes:zuckerberg:so boring	0.1462	
vr:the australian public's attitude:exploring:yuri_librarian it's cool	0.1462	
vr:atari cofounder nolan bushnell:startups:virtual reality	0.1103	
Storyline	Classifier weight	
vr:la france:le grand cran:vive	2.0841	
vr:1:11:us a mobile monday detroit	1.9923	
vr:1:2016:titres playstation vr annoncés	1.980	

Table 9. Top named entities in the data elements for top storylines are shown in Table 10.

Besides well known companies in Table 10, the bulk of named entity list comprises of new companies such as 'Unimersiv - VR Education', 'Lunar Flight CV' and 'opTic' that are of specific interest to investors. The utility of the virtual reality and augmented reality results were more mixed compared to the connected car results. Themes resulting from topic modeling varied from Lenovo's use of Google's Tango augmented reality technology in their new smartphones to robbers using the mobile game Pokemon Go to find and lure unsuspecting

Table 10. AR top named entities by info gain feature selection, topic modeling and classifier weights.

Organization	Topic frequency	Person	Topic frequency
HTC Vive	40	VR	46
Sony	12	Unity	10
Microsoft	13	David	4
Organization	Info gain frequency	Person	Info gain frequency
HTC Vive	28	des	9
Sony	12	phil	6
Microsoft	9	lucas	7
Organization	Classifier frequency	Person	Classifier frequency
HTC Vive	47	VR Shell	25
Google	45	Kevin Durant	7
Samsung	35	vida	7
Nintendo	9	una	9

victims, and the impending launch of Samsung Gear VR 2, Samsung's latest virtual reality headset. Storylines identified thru information gain and classification included publicity for mobile Monday Detroit, an event showcasing AR and VR demos, Mark Zuckerberg's announcement of Open Cellular which is tangentially related to the topic of virtual reality, and news around the development of Sony's Playstation VR gaming platform. The organizations identified by the algorithms included some of the biggest names in virtual reality such as Google, HTC, Sony, and Samsung as well as other organizations less known for their efforts including Nintendo and Microsoft. Nintendo was making a huge splash in the market with its wildly successful Pokemon Go augmented reality mobile game and this analysis perhaps acted as an early indicator of how successful the game would eventually become. Meanwhile, Microsoft was making news for using virtual reality for autistic kids in their preparation for jobs. The identified persons included Phil (Phil Spencer) the Xbox chief at Microsoft, David (David Joyner) of Georgia Tech who publishes a blog summarizing news and events in the area of virtual reality, and Kevin Durant, professional basketball player, who was making headlines as the Golden State Warriors used virtual reality to recruit Durant. As was the case with connected car, some of the identified names were actually organizations or other entity types. One such example was Unity, a game development platform used by leading game developers. Overall, the algorithms effectively managed to refine a large volume of raw data (over 140,000 virtual reality related tweets and nearly 40,000 connected car related tweets) to a much more manageable and targeted dataset for the analyst to examine and further research. Furthermore, it would be interesting to track the results of multiple days' worth of data to see what results would emerge.

5.3   Performance

The performance of the techniques used in event creation at different levels of distribution is evaluated in this subsection. The results for running the techniques on various sized clusters are presented. The experiments were run on AWS using Elastic MapReduce clusters running Spark. This allows for clusters to be configured on demand on the cloud so that scalability of the techniques on different sized datasets and clusters can be tested. Cluster nodes are of type m3.2xlarge with 8 vCPU processors and 30 GB of RAM.

Figure 5 shows the performance of topic modeling on various sized clusters. The same code run on a single node is an approximation of how similarly written single node sequential version will perform. The results show clearly that with increasing number of storylines, the time taken to perform topic modeling on the storylines does not increase significantly on an 8 node cluster but continues to increase for sequential runs. Beyond a dataset of certain size the single node execution generates out of memory errors. Topic modeling is highly iterative hence its distribution is critical to its being able to scale to larger datasets. Results are similar for Information gain based features selection and SVM modeling executions on multiple sized training datasets.

Figure 6 shows the performance of spatio-temporal entity identification. The results clearly show that the process of identifying spatial and temporal entities is highly parallelizable with testing each storyline against GATE API independent of others. Figure 8 shows storylines categorization performance using feature selection generated information gain weights. These results show that once feature selection has generated top info gain entities, categorizing storylines under those events is highly parallelizable and scalable with running times staying stable with increasing data and cluster sizes. Figure 7 shows results for categorizing storylines into events using topic modeling weights. This was done for 3 topics

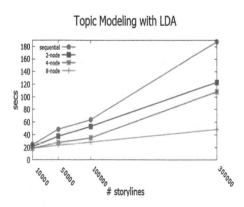

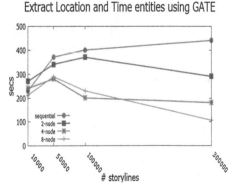

Fig. 5. Performance of topic modeling on various cluster and storyline data sizes.

Fig. 6. Performance of spatio-temporal entity creation on various cluster and storyline data sizes

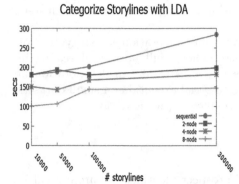

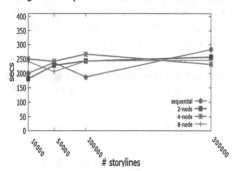

Fig. 7. Performance of storylines categorization into events generated from topic modeling on various cluster and storyline data sizes

Fig. 8. Performance of storylines categorization into events generated from feature selection on various cluster and storyline data sizes

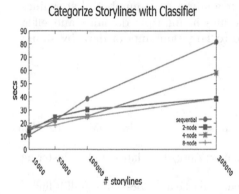

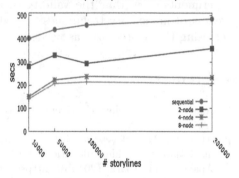

Fig. 9. Performance of storylines categorization based on svm model scoring on various cluster and storyline data sizes.

Fig. 10. Performance of extracting top people and organizations from tweets of top storylines as identified by FS, LDA and SVM on various cluster and storyline data sizes.

and each storyline was tested against multiple events, yet the process is highly scalable and parallelizable.

Figure 9 shows the performance of scoring storylines using SVM classifier. This is also highly parallelizable as once a model is built, it can score storylines independent of each other. Figure 10 shows the performance of top named entities as organizations and people extraction to identify top prospects from tweets of top storylines. The performance over increasing larger test datasets and clusters shows that this is highly scalable.

The results clearly show the scaling of events generation for storylines for large datasets. Increasing the size of cluster allows full horizontal scaling in DISTL. Increased overhead of Spark in some cases results in deterioration in performance on small clusters as compared to serial execution on small datasets but with larger sets of storylines the performance improves vastly.

6 Conclusions

Event based storylines categorization and prospect identification are critical in providing analysts meaningful ways to analyze large number of storylines generated from social and open data. Proposed framework DISTL is effective in generating events and identifying prospects at a large scale. The supervised and unsupervised learning techniques incorporate domain expert knowledge and generate events and prospects from storylines. Rules allow flexibility in categorizing storylines under events. The resulting events incorporate theme, location and time and are useful in allowing analyst analyze large number of storylines. Experiments show that the validity of results in multiple domains and effectiveness of in-memory techniques to scale to large amounts of data by simply increasing the cluster sizes as needed.

References

1. Agarwal, C., Subbian, K.: Event detection in social streams. In: SDM, pp. 624–635 (2012)
2. Apache and Spark (2015a). https://spark.apache.org/docs/latest/mllib-clustering. html#latent-dirichlet-allocation-lda
3. Apache and Spark (2015b). https://spark.apache.org/docs/latest/mllib-linear-methods.html\#inear-support-vector-machines-svms
4. Apache and Spark. Spark programming guide (2015c). http://spark.apache.org/ docs/latest/programming-guide.html
5. Apache, Spark, and Packages (2015). https://github.com/wxhc3sc6opm8m1hxbomy /spark-mrmr-feature-selection
6. Chae, J., Thom, D., Bosch, H., Jang, Y., Maciejewski, R., Ebert, D., Ertl, T.: Spatiotemporal social media analytics for abnormal event detection and examination using seasonal-trend decomposition. In: 2012 IEEE Conference on Visual Analytics Science and Technology (VAST), pp. 143–152 (2012)
7. Chen, F., Neill, D.B.: Non-parametric scan statistics for event detection and forecasting in heterogeneous social media graph. In: ACM SIGKDD, pp. 1166–1175 (2014)
8. Cunningham, H., Maynard, D., Bontcheva, K., Tablan, V.: Developing language processing components with gate version 8. University of Sheffield Department of Computer Science (2014)
9. Dean, J., Ghemawat, S.: Mapreduce: simplified data processing on large clusters. Commun. ACM **51**(1), 107–113 (2008)

10. Weng, J., Lee, B.-S.: Event detection in twitter, pp. 401–408. AAAI (2011)
11. Keskisärkkä, R., Blomqvist, E.: Semantic complex event processing for social media monitoring-a survey. In: Proceedings of Social Media and Linked Data for Emergency Response (SMILE) Co-Located with the 10th Extended Semantic Web Conference, Montpellier, France, CEUR Workshop Proceedings (2013)
12. Lappas, T., Vieira, M.R., Gunopulos, D., Tsotras, V.J.: On the spatiotemporal burstiness of terms. Proc. VLDB Endow. **5**(9), 836–847 (2012)
13. Leetaru, K., Schrodt, P.A.: GDELT: global database of events, language, and tone. In: ISA Annual Convention (2013)
14. Li, C., Sun, A., Datta, A.: Twevent: segment-based event detection from tweets. In: Conference on Information and Knowledge Management, pp. 155–164 (2012a)
15. Li, R., Lei, K.H., Khadiwala, R., Chang, K.: Tedas: a twitter-based event detection and analysis system. In: Proceedings of 28th IEEE Conference on Data Engineering (ICDE), pp. 1273–1276 (2012b)
16. Petrovic, S., Osborne, M., McCreadie, R., Macdonald, C., Ounis, I., Shrimpton, L.: Can twitter replace newswire for breaking news? In: 7th International AAAI Conference on Weblogs and Social Media (ICWSM) (2013)
17. Pohl, D., Bouchachia, A., Hellwagner, H.: Automatic sub-event detection in emergency management using social media. In: Proceedings of the 21st International Conference Companion on World Wide Web (WWW 2012 Companion), pp. 683–686, New York, NY, USA. ACM (2012)
18. Radinsky, K., Horvitz, E.: Mining the web to predict future events. In: WSDM, pp. 255–264 (2013)
19. Reuter, T., Buza, L.D.K., Schmidt-Thieme, L.: Scalable event-based clustering of social media via record linkage techniques. In: ICWSM (2011)
20. Saha, A., Sindhwani, V.: Learning evolving and emerging topics in social media: a dynamic NMF approach with temporal regularization. In: Proceedings of the Fifth ACM International Conference on Web Search and Data Mining (WSDM 2012), pp. 693–702, New York, NY, USA. ACM (2012)
21. Sakaki, T., Okazaki, M., Matsuo, Y.: Earthquake shakes twitter users: real-time event detection by social sensors. In: WWW, pp. 851–860 (2010)
22. Shukla, M., Santos, R.D., Chen, F., Lu, C.-T.: Discrn: a distributed storytelling framework for intelligence analysis. Virginia Tech Computer Science Technical report (2015)
23. Walther, M., Kaisser, M.: Geo-spatial event detection in the twitter stream. In: Serdyukov, P., Braslavski, P., Kuznetsov, S.O., Kamps, J., Rüger, S., Agichtein, E., Segalovich, I., Yilmaz, E. (eds.) ECIR 2013. LNCS, vol. 7814, pp. 356–367. Springer, Heidelberg (2013). doi:10.1007/978-3-642-36973-5_30
24. Zaharia, M., Chowdhury, M., Das, T., Dave, A., Ma, J., McCauly, M., Franklin, M.J., Shenker, S., Stoica, I.: Resilient distributed datasets: a fault-tolerant abstraction for in-memory cluster computing. In: Presented as Part of the 9th USENIX Symposium on Networked Systems Design and Implementation (NSDI 2012), pp. 15–28, San Jose, CA. USENIX (2012)
25. Zhao, L., Chen, F., Dai, J., Lu, C.-T., Ramakrishnan, N.: Unsupervised spatial events detection in targeted domains with applications to civil unrest modeling. PLOS One **9**(10), e110206 (2014)
26. Zhao, L., Chen, F., Lu, C.-T., Ramakishnan, N.: Spatiotemporal event forecasting in social media. In: SDM, pp. 963–971 (2015)
27. Zhou, X., Chen, L.: Event detection over twitter social media streams. VLDB J. **23**(3), 381–400 (2014)

28. Li, J., Tsang, E.P.K.: Investment decision making using FGP: a case study. In: Proceedings of the 1999 Congress on Evolutionary Computation (CEC 1999), vol. 2, no. 2, pp. 1259–1279 (1999)
29. Shukla, M., Santos, R.D., Fong, A., Lu, C.-T.: DISTL: distributed in-memory spatio-temporal event-based storyline categorization platform in social media. In: Proceedings of the 2nd International Conference on Geographical Information Systems Theory, Applications and Management, pp. 39–50, Italy, Rome (2016)
30. Cheh, J.J., Weinberg, R.S., Yook, K.C.: An application of an artificial neural network investment system to predict takeover targets. J. Appl. Bus. Res. (JABR) 15(4), 33–46 (2013)
31. Geum, Y., Lee, S., Yoon, B., Park, Y.: Identifying and evaluating strategic partners for collaborative R&D: index-based approach using patents and publications. Technovation 33(6), 211–224 (2013)

Automated Walkable Area Segmentation from Aerial Images for Evacuation Simulation

Fabian Schenk[✉], Matthias Rüther, and Horst Bischof

Institute for Computer Graphics and Vision,
Graz University of Technology, Graz, Austria
{schenk,ruether,bischof}@icg.tugraz.at

Abstract. In this paper, we propose a novel, efficient and fast method to extract the walkable area from high-resolution aerial images for the purpose of computer-aided evacuation simulation for major public events. Compared to previous work, where authors only extracted roads and streets or solely focused on indoor scenarios, we present an approach to fully segment the walkable area of large outdoor environments. We address this challenge by modeling human movements in the terrain with a sophisticated seeded region growing algorithm (SRG), which utilizes digital surface models, true-orthophotos and inclination maps computed from aerial images. Further, we propose a novel annotation and scoring scheme especially developed for assessing the quality of the extracted evacuation maps. Finally, we present an extensive quantitative and qualitative evaluation, where we show the feasibility of our approach by evaluating different combinations of SRG methods and parameter settings on several real-world scenarios.

Keywords: Aerial images · Walkable area · Seeded region growing · Evacuation maps · Accessibility

1 Introduction

Millions of people visit sports competitions, concerts and religious celebrations every year and with the ever growing population the number is not likely to decrease. There is a large variety of possible emergencies (natural disasters, fire, terrorist attacks, bombings) with most of them requiring a full or partial evacuation of the event area. Adequate safety measures are required to prevent crowd disasters like the one at the Love Parade 2010 (Duisburg, Germany) [1,2], where 21 people died and more than 500 were injured or the tragic incident that occurred during the Water Festival 2010 (Phnom Penh, Cambodia) [3] with around 380 dead.

To assure the safety of major public events, evacuation simulation is an important preliminary step in the planning stage, but is normally a tedious and time-consuming task due to the complex layout of large event sites. For computer-aided evacuation simulation and planning for outdoor events, digital

© Springer International Publishing AG 2017
C. Grueau et al. (Eds.): GISTAM 2016, CCIS 741, pp. 89–108, 2017.
DOI: 10.1007/978-3-319-62618-5_6

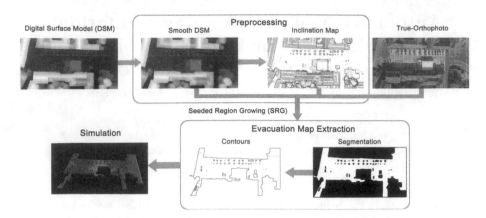

Fig. 1. We use a sophisticated seeded region growing (SRG) algorithm to segment the walkable area (WA) incorporating information from the DSM, true-orthophoto and orientation map. From this segmentation, the contours can be extracted and used in most common evacuation simulation programs.

maps are required. A computer can then perform a wide variety of simulations on the given topology using different hazards, escape routes and human properties like walking speed or age.

The technology for recording aerial images has been around for some time, but in recent years high-resolution cameras have improved the accuracy significantly and small unmanned aerial vehicles (UAV) can perform on-demand recordings of certain areas. In previous work on outdoor evacuation simulations, [4,5] only extracted streets and roads, but did not achieve very high accuracy. It is not enough to simply segment a certain height level, roads or just flat areas because humans can walk on different surfaces (slopes, stairs). If measurements are wrong or blocking structures are missing in the digital map, a simulation cannot be performed accurately.

To our knowledge, we are the first to address the difficult challenge of getting an accurate digital map of the WA for evacuation planning in outdoor environments using high-resolution aerial images, which is a complex segmentation problem comprising the following challenges:

- Incorporating measurements correctly into the map
- Providing accurate contours for buildings and blocking structures (walls, food stands, tents,...)
- Segmenting potential emergency exits (roads,...)

Extending work presented by Schenk et al. [6], we introduce a novel, efficient and easy-to-use approach to generate highly accurate digital maps of the WA from high-resolution aerial images. Our approach generalizes very well to various scenarios and cities and by adapting the parameters for slopes and stairs we can also incorporate the special needs of handicapped people (wheelchair users, elderly persons). These maps can then be used for evacuation planning and simulation (see Fig. 1).

Further, we present a new annotation scheme to assess the quality of the extracted maps with regard to evacuation simulation. Additionally, aerial images give us the opportunity to acquire data shortly before the actual event and to include structures like stages, food stands and tents into our map.

The paper is organized as follows. In Sect. 2, we discuss related work on evacuation simulation and street segmentation and its limitations. Section 3 describes the input data, the necessary preprocessing steps and finally our method to segment the WA. An exhaustive qualitative and quantitative evaluation of different SRG algorithms and their combinations as well as various parameter settings is shown in Sect. 4. Section 5 concludes the paper and gives some ideas about future work on this topic. In the Appendix, we will show an evacuation simulation using our extracted digital maps.

2   Related Work

Evacuation simulation is an active research area and many different software tools have been developed, which can be classified as microscopic and macroscopic. In microscopic models [7–9], space, time and persons are represented on a very fine scale, while macroscopic approaches [10] use the analogy with the flow of liquids. All tools require a digital map for evacuation simulation and most of them support CAD (computer-aided drafting) models. Previous work can be divided into indoor and outdoor scenarios, with the main difference being that for the latter usually no exact digital maps of the WA are available.

Indoor scenarios have been extensively studied in recent years and normally CAD and sometimes even 3D models are provided by architects from which the WA can be extracted directly. Johnson et al. [11] performed an evacuation simulation to improve security and safety of the 2012 Olympic venues. Shi et al. [12] derived an engineering formula for the calculation of the evacuation time by extensively studying different emergency scenarios in the very crowded metro stations of Tokyo, Japan. An indoor fire simulation model was presented by Tang et al. [13], which is based on fire fields, human behavior and building geometry. In [9], various evacuation strategies for the International Terminal at Los Angeles International Airport were modeled and studied.

Outdoor Scenarios have been analyzed with the main focus on developing evacuation strategies for tsunami incidents. In this context, Mas et al. [14] proposed an evacuation model with a tsunami simulation for casualty estimation for the urban area of La Punta, Peru. Evacuation analysis at city-scale by the example of Padang, Indonesia for the case of a tsunami incident was performed in [4], which was later extended by a simulation framework [5]. In their simulations, the authors achieved building level accuracy by extracting semantic labels from four band satellite images. With their city-wide analysis they do not achieve very high accuracy and for planning public events only small areas of a city like squares or parks are of particular interest.

Koch et al. [15] introduced a GIS-based Incident Management Preparedness and Coordination Toolkit (IMPACT) to help first responders in various emergencies. With this software, a user can quickly draw enclosing boundaries and hazards around an interesting area in an aerial image to perform evacuation simulation. In [8], the authors performed crowd simulations for large outdoor events by the examples of the World Youth Day 2005 in Cologne and an egress (non-emergency) from a football stadium but mainly focused on evacuation details (reaction time, walking speed).

Road and street segmentation from aerial images acquired by satellites or UAVs is also a well researched topic [16–19]. Even though these approaches give us an idea of the WA, humans can access more than just roads and streets and for evacuation planning a more sophisticated segmentation is needed. OpenStreetMap provides geospatial data for certain urban areas, but streets are mostly modeled only as a line, thus the actual width is unknown and other annotations of blocking structures or the extents of squares, are often not very not accurate or outdated.

Land-use classification from aerial images has been presented in [20,21], while Han et al. [22] try to find objects and their respective bounding boxes. These methods are well suited for land-use study of whole cities or even larger regions but lack the necessary accuracy for WA analysis, where the contours of blocking structures like buildings are essential. The number of object classes is usually limited, thus misclassification in the final result are very common.

In contrast to previous approaches, we are not limited by object classes and provide a complete segmentation of the walkable area of large outdoor environments.

3   Evacuation Map Extraction

The main challenge is to get an accurate digital map of the WA from high-resolution aerial images (see Fig. 1). From these images we can calculate the corresponding digital surface models (DSMs) and then we use edge-preserving smoothing to facilitate the subsequent calculation of the inclination map (see Sect. 3.1).

One downside of the very high resolution is the large amount of data and processing a whole city can take from hours to days, depending on the available computational power. Thus, we perform our segmentation in a region of interest (ROI). Within this ROI the user then has to choose a point manually. Starting from this point a seeded region growing (SRG) [23] algorithm segments the WA (see Sect. 3.2). The result is a binary segmentation of the WA, which can be exported as CAD model to most common evacuation simulation program (see Sect. 3.3).

3.1   Aerial Image Input Data and Preprocessing

The acquisition of the aerial images and calculation of the DSMs and true-orthophotos are not part of our approach but will be briefly explained. Aerial images are typically recorded with three (RGB) or four channels (RGB + near infrared) by UAVs like drones or smaller airplanes because of the higher spatial resolution compared to satellites. UAVs follow a certain pattern when taking images of an area to get multiple, overlapping views of all the structures. Then the high-resolution aerial images are used for triangulation, which is also known as Structure from Motion (SfM), to get accurate photo alignments from image measurements only [24]. For SfM, correspondences between images have to be established, which is done by using Scale Invariant Feature Transform (SIFT) features with Lowe distance ratio [25] followed by RANSAC [26] outlier filtering. Then all camera poses are estimated from these correspondences with sparse bundle adjustment [27]. The result can be further refined using the GPS signal and known ground control points.

A dense height map for each image is estimated with a multi-view reconstruction approach similar to the one presented in [24,28]. The basic idea is that the actual height value of a pixel can be found by comparing two overlapping images from different views. A plane sweep approach [29] then shifts one image in a certain direction, while the other one stays fixed and calculates a matching cost between these two, resulting in a 3D cost volume. In [28], they use a winner-takes-all method on the cost volume and always choose the cheapest matching cost for each pixel as the correct height. Alternatively, an optimizer on the cost volume followed by multi-depth filtering can be used to get the correct height values.

The next step is generating a DSM I, which represents the height information of Earth's surface including all objects (buildings, trees, cars,...). The DSM is only a 2.5D model because the aerial images are recorded from above, where only part of the surface is visible. Recording what is beneath is not possible (e.g., a tunnel under a mountain or a river under a bridge) and therefore only the top surface of a structure is present in the DSM. From the overlapping height maps calculated in the previous step we get multiple proposals for the height of a pixel. For each pixel the most likely height value from all the proposals is found and incorporated into the DSM. The final height resolution of the DSM is usually much higher than that of the overlapping height maps because of the multiple measurements.

By back-projecting the point from the DSM into the camera and coloring the pixel accordingly a true-orthophoto with a uniform scale (like an ordinary map) can be generated and use for measuring distances. The applications of DSMs and true-orthophotos are widespread and include infrastructure planning, 3D building reconstruction, city modeling and simulations for natural disaster management. As input for our algorithm we use the DSM and the corresponding true-orthophoto (see Fig. 2).

We are not only interested in flat areas but also in ramps with a moderately steep slope and stairs, which have a larger difference in height (around 16–20 cm). To get an idea about the inclination of a surface, we have to calculate the surface normal representation from the DSM. The DSM only has one channel (height information) but for the calculation of the surface normal representation we need a 3D representation. Therefore, we generate such a 3D representation I_{3D} of I_S, where $I_{3D}(x, y)$ is given as

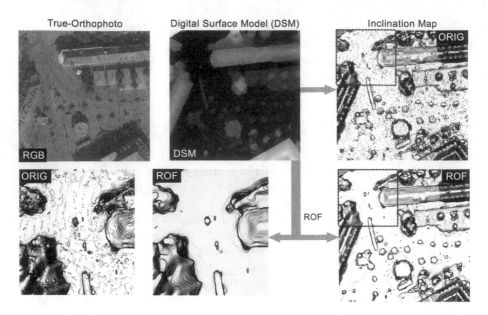

Fig. 2. For the SRG segmentation we use a true-orthophoto, a DSM and an inclination map. In the DSM model, height is represented as intensity values (higher is brighter). Calculating the inclination map directly from the DSM shows a stair-casing effect (depicted as ORIG), which is greatly reduced when computed from a smooth DSM (shown as ROF). Images are contrast-enhanced for visualization purposes.

$$\boldsymbol{I}_{3D}(x, y) = (x \cdot s_x, y \cdot s_y, I_S(x, y)), \tag{1}$$

with s_x and s_y representing the spatial resolution in x- and y-direction. Typically, we can assume $s_x = s_y$.

The basic idea for the surface normal calculation is to first compute the two tangential vectors $\boldsymbol{t}_x$ and $\boldsymbol{t}_y$ at each position (x, y) from its 4-neighborhood (see Fig. 3(a)).

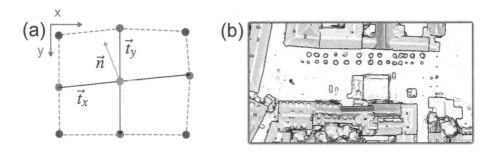

Fig. 3. (a) depicts the normal vector $\boldsymbol{n}$ and the tangential vectors $\boldsymbol{t}_x$ and $\boldsymbol{t}_y$ computed from the 4-neighborhood. (b) shows the inclination of the surface as $cos(a)$. White surfaces are flat, while darker ones represent an inclination.

We define t_x and t_y as:

$$t_x = I_{3D}(x+1, y) - I_{3D}(x-1, y), \tag{2}$$

$$t_y = I_{3D}(x, y+1) - I_{3D}(x, y-1). \tag{3}$$

We achieve fast computation and minimal memory access with integral images as described in [30]. At position (x, y), the surface normal vector n is then given as:

$$n = t_x \times t_y. \tag{4}$$

For the next steps, the inclination of the slope is important, which is the angle α between a view vector v and the surface normal vector n. To be in the range of $[0, 1]$, we choose $|cos(\alpha)|$ as representation of the inclination, which is given by the definition of the dot product as:

$$cos(\alpha) = \frac{\langle v, n \rangle}{|v||n|}. \tag{5}$$

In our DSMs, the camera typically has a top-down view of the scene, thus we assume a view vector $v = (0, 0, 1)$ with $|v| = 1$. If we compute the dot product, the equation simplifies to the normalized z-component of the surface normal vector n:

$$cos(\alpha) = \frac{n_z}{|n|}. \tag{6}$$

A flat surface has an angle $\alpha = 0°$, thus $cos(\alpha) = 1$, i.e. white in the inclination map (see Fig. 3(b)). The 4-neighborhood is used for the calculation of the inclination, which results in an angle between the view vector v and n at edges (see Fig. 6(c)).

Alternatively, one could compute a similar map by using an edge weighting term $e^{-\beta|\nabla I_S|}$, where I_S is the smooth DSM model and β is a constant.

As depicted in Fig. 2, calculating the surface normals directly from the DSM computed from the aerial images leads to artifacts due to the plane-sweep approach [29]. We address this problem by using a smoothing algorithm, which smooths the DSM I while keeping the edges intact. Edge information is crucial because we want to keep accurate contours of buildings and other structures. The difference between calculating this representation on the smooth and non-smooth DSM can be seen in Fig. 2. For our particular problem we choose the model introduced by Rudin, Osher and Fatemi (ROF) [31]:

$$\min_{I_S} \left\{ \int_\Omega |\nabla I_S| d\Omega + \frac{\lambda}{2} \int_\Omega (I_S - I)^2 d\Omega \right\}, \tag{7}$$

where the first part is the regularization term, the second the data fidelity term, I_S a convex, continuous function representing the smooth image and Ω the image space. We want to reconstruct an image I_S, which is smooth but also similar to the original input image I. In order to get a smooth I_S, the regularization term reduces the differences within I_S by penalizing the L1-norm of the gradients. The data fidelity induces similarity by punishing differences between I_S and I with a quadratic norm. The weighting term λ serves as a trade-off between data fidelity and regularization. A higher λ gives more emphasize to the data fidelity term, leading to I_S being more similar to the original I while a lower λ results in a very smooth I_S. Figure 4 depicts the influence of different λ values and shows that with a very low λ, e.g. $\lambda = 1$, most of the edges are lost, while higher values preserve the edges. We solve this convex optimization problem with a primal-dual optimization algorithm [32]. After denoising, the results look smoother and are more suitable for our approach (see Fig. 2, ROF).

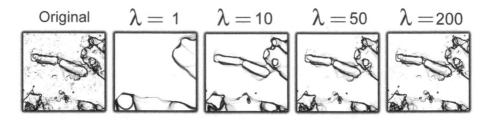

Fig. 4. This figure depicts the influence of the trade-off parameter λ, with a low value smoothing most of the structures, while a higher value keeping more edges.

3.2   Segmentation of the Walkable Area (WA)

The main and most crucial part in our approach is the extraction of the WA. For an accurate simulation it is not enough to just segment a certain height level because the whole WA including slopes and stairs is interesting. This is a complex search problem because accessibility highly depends on the topology of the surface. An area might be inaccessible from one side due to a large height difference but by using stairs or ramps on another side it might become accessible, thus we have to account for such a topology. As a result we want to have a binary segmentation, where all points accessible by humans from the chosen starting point are labeled 1 and the rest 0.

For our segmentation approach, we assume a constrained set of four possible walking directions for a person, for-/backward and left/right. In a top-down view, this is represented by the 4-neighborhood. We decided to use a seeded region growing (SRG), which models these movements as it starts from a manually chosen seed pixel and adds adjacent ones to the segmentation if they fulfill certain conditions (see Fig. 5). They then in turn become the next seeds.

This method is especially feasible for our problem because it can model walking directions very well and we can easily incorporate different segmentation conditions for various input data. We utilize all the available information by using the DSM, the true-orthophoto and the inclination map as depicted in Fig. 1.

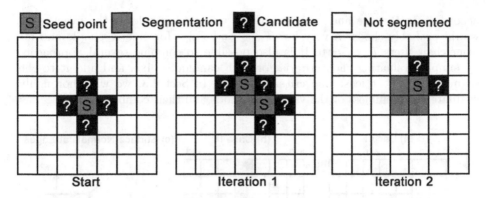

Fig. 5. Here, the basic idea of a SRG algorithm is depicted. Starting from a seed pixel S, adjacent pixels are added to the segmentation if they fulfill a certain condition and in turn become the next seeds. This process continues till no more pixels can be added to the segmentation.

With our approach we want to segment the WA including slopes and stairs. Starting from a center point p_c we have the two different cases presented in Fig. 6(a,b).

Case I. In this case, we check the pixels in the 4-neighborhood (see Fig. 6(a)). We calculate the height difference Δh between a center point p_c and its surrounding p_{1-4} from the DSM I_S and add the respective point if it Δh is smaller than a threshold value T_{slope}. The true-orthophoto gives us additional color information and we assume that in most cases an accessible area with the same color does not suddenly become inaccessible. Thus, we additionally allow a slightly greater height difference Δh of $2 \cdot \Delta T_{slope}$ when the color difference Δc is below a threshold T_{color}. We use I_S instead of the inclination map because of the clearer edges (see Fig. 6(c)).

Case II. Stairs usually have a higher height difference Δh but are still walkable by humans. To include such a concept in our algorithm, we allow growing over a greater height difference T_{stair} if the orientation is nearly horizontal/flat ($cos(\alpha) \geq 1 - \epsilon_s$). We have to check the neighborhood $(x \pm 3, y)$ and $(x, y \pm 3)$ because the surface normal vectors are not pointing upwards at edges (see Fig. 6(b, c)). If a point p_n fulfills the criteria, we also add the points between it and p_c (depicted in gray) to the segmentation.

We can divide our SRG approach into the three different parts $SRG_{\widetilde{SL}}$, SRG_{ST} and SRG_C, depending on the segmentation criteria:

- $SRG_{SL} : \Delta h \leq T_{slope}$.
- $SRG_C : \Delta h \leq 2 \cdot T_{slope}, \Delta c \leq T_{color}$.
- $SRG_{ST} : \Delta h \leq T_{stair}, cos(\alpha_{p_c}), cos(\alpha_{p_n}) \geq 1 - \epsilon_s$.

with $cos(\alpha_{p_c})$ and $cos(\alpha_{p_n})$ the inclination at the center point p_c and the one to be added p_n. ϵ_s and T_{color} are constants, while T_{slope} and T_{stair} can be adapted according to the age and mobility of the expected people at the event. One would probably choose a very low T_{slope} and T_{stair} for elderly people and for wheelchair users $T_{stair} = 0$. It is also possible to combine several approaches by simply evaluating each criteria for each candidate pixel and add it to the segmentation if at least one criteria is true. In Sect. 4.3, we will thoroughly evaluate the usefulness of the various methods and their combinations.

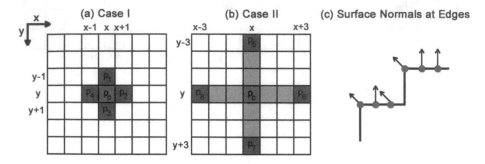

Fig. 6. (a) and (b) are the two different neighborhoods (red) around the center point p_c (cyan) considered during the SRG algorithm. Case I is the 4-neighborhood (p_{1-4}), used for growing over a slope or color difference. Case II shows a larger neighborhood (p_{5-8}) for growing over stairs, which is necessary because of the surface normal orientation at edges depicted in (c) (Color figure online)

Slopes, and stairs both introduce the same kind of problem. When structures of different height levels are next to each other (no pixels between them) then a complete segmentation would lead to a whole area being segmented, even though it is not possible to go from one height level to the other (see Fig. 7(a)). Usually, height information cannot be used in the simulation programs, thus we put a barrier with width of one pixel between the structures if their height difference $\Delta h > T_{stair}$. Figures 7(b, c) depict the segmentations and possible walking directions (red arrows) with and without barrier, demonstrating the feasibility of this approach. The result is a complete, binary segmentation of the

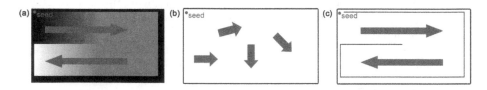

Fig. 7. (a) is an artificially generated DSM with two adjacent ramps and the seed point at the top left corner. (b) shows the segmentation when not taking the height difference into account, while (c) is the correct segmentation with barriers. The possible walking directions are depicted as red arrows. (Color figure online)

WA including ramps and stairs. Different digital maps of the WA extracted with our method and the corresponding ground truth will be presented in Sect. 4.3.

3.3 Exporting the CAD Model

Normally, evacuation simulation tools cannot directly use a binary segmentation but require a computer-aided drafting (CAD) model. For the CAD model only the enclosing boundaries and the contours of blocking structures on the inside, which are represented as holes in the segmentation, are of particular interest. We extract the boundaries and then use them to generate the CAD model for most common simulation tools. In Appendix A, we show how we can utilize our binary segmentation in the evacuation simulation tool PedGo [8].

4 Experiments and Results

In this section, we show an extensive evaluation of the different SRG methods presented in Sect. 3.2 on real-world scenarios and demonstrate the feasibility of our approach. We further propose a novel annotation scheme for assessing the quality of extracted evacuation maps (see Sec. 4.1). In the last part, we study different parameter settings for T_{stair}, λ and T_{slope} and analyze their influence on the segmentation.

4.1 Experimental Setup

For evaluation of computer vision methods the results are usually compared to known and more accurate references, which are commonly referred to as ground truth (GT). Getting any kind of GT data for aerial images is very difficult and to our knowledge no ground truth data is available for the special case of evacuation maps. Therefore, a computer vision expert manually annotated the WA based on the true-orthophotos. Accurate manual segmentation is also challenging because the true-orthophotos are calculated from multiple recordings and often exact borders are hard to determine.

For the evaluation of our SRG algorithms, we choose four outdoor scenarios from various cities and with different spatial resolutions. We use the Dice coefficient (DC) [33] and the Jaccard similarity (JS, also known as Tannimoto Coefficient) [34] to compare our segmentation SEG to the GT. $|SEG|$ and $|GT|$ denote the sum of segmented pixels.

Dice Coefficient (DC)
The DC is a measure for comparing 2D region overlap with a range from $[0, 1]$ and is defined as:

$$DC = \frac{2|SEG \cap GT|}{|SEG| + |GT|} \tag{8}$$

Jaccard Similarity (JS)
The JS is very similar to the DC and ranges from $[0, 1]$ but is usually lower. It is defined as:

$$JS = \frac{|SEG \cap GT|}{|SEG \cup GT|} \tag{9}$$

The DC and JS are good indicators for the overall segmentation quality but do not give us a measurement for usefulness of the extracted evacuation map. Further, the segmented areas are quite large ($>1MP$) and missing small but important structures hardly affects the overall score.

Evacuation Map Annotation
To assess the quality of extracted evacuation maps, we propose a new annotation scheme with two types of labels:

- Potential evacuation routes GT_P
- Structures that must not be segmented GT_B

These additional GT annotations can then be compared to the segmentation. We define two scores S_P, S_B in the range $[0, 1]$ to assess the segmentation quality of potential evacuation routes and blocking structures. They are defined as:

$$S_P = \frac{|SEG \cap GT_P|}{|GT_P|}, \tag{10}$$

and

$$S_B = 1 - \frac{|SEG \cap GT_B|}{|GT_B|}, \tag{11}$$

with GT_P as the annotated potential evacuation routes, GT_B the blocking structures and SEG the segmentation result. These two values must always be evaluated jointly because $S_P = 100\%$ if the whole image is segmented, while $S_B = 100\%$ if nothing is segmented. Examples for such annotations can be seen in Fig. 8 (c, f, i, l) with GT in white, GT_B in red and GT_P in green.

All the computations were performed on an Intel i7-4790 (3,6 GHz × 8) with 32 GB of memory running Ubuntu Linux 14.04. The denoising and SRG algorithms are completely implemented in C++ using OpenCV 3.0.0 to allow fast computations. For a typical 15 MP image, denoising is by far the most computationally expensive operation (around 2 min), while segmentation works a lot faster (<1 s).

4.2   Aerial Image Data

Real-world scenarios are very challenging because the algorithm has to deal with vegetation, different regional building styles, cars and other obstacles. The performance of our method mainly depends on the quality of the DSMs and true-orthophotos, which are usually subject to noise and reconstruction artifacts. In all the real-world experiments, the GT is compared to the automatically generated results from the algorithm. We choose four sites from three different cities (**Exp. A–D**).

Experiment A is the Marienhof, which is a big park with space for a lot of people and many adjacent streets (see Fig. 8(a)). In this recording only gray-scale true-orthophotos were available.

Experiment B is the Jakominiplatz, which has a very difficult setup due to the many bus stops, buses, trams and street lamps (see Fig. 8(d)).

Experiment C is the Karmeliterplatz, where an event took place at the moment of recording (see Fig. 8(g)). A large tent with a stage in front is present.

Experiment D shows a bridge over a small road next to the Expo Plaza in Hannover, Germany (see Fig. 8(j)). This is the only drone recording and therefore has a much higher spatial resolution. The setup is especially interesting because it includes stairs, slopes, cars, buildings and different height levels (bridge, road).

The spatial resolution of **Experiment A–C** is $s_x = s_y = 10$ cm, while in **Experiment D** it is $s_x = s_y = 3$ cm.

4.3   Results and Discussion

In this section, we present an extensive quantitative and qualitative evaluation of the segmentation results for **Experiment A–D**, followed by a discussion, where we also investigate various parameter settings for T_{stair}, T_{slope} and λ.

For the evaluation, we study the different SRG methods presented in Sect. 3.2 and their combinations. We present the resulting scores for DC, JS and S_P as percentages in Table 1. In our evaluation we only included results, where all the blocking structures were not segmented ($S_B = 100\%$). We assure a fair comparison by optimizing the parameters T_{slope} (up to a maximum of 35% of the spatial resolution $s_{x,y}$) and λ (up to a value of 500) for each SRG method and their combinations. Further, T_{stair} was set to the realistic values 10 and 20 cm, T_{color} was set to ±3 for each channel (R,G,B) and $\epsilon_s = cos(10°)$. For **Experiment A** only gray-scale true-orthophotos were available, thus we simply used the one intensity channel for each of the three color channels (R, G, B). The settings for T_{stair}, T_{slope} and λ for each experiment are presented in Table 2 and their choice will be discussed later.

Table 1 shows that the worst scores were achieved by SRG_C, which cannot handle color changes (i.e., shadows) and usually only segments an area surrounding the seed point. SRG_{ST} performs far better but can only segment areas, which are either flat or stairs. SRG_{SL} shows high scores but is in general outperformed by combinations of different SRG methods. Overall, the combination of the three SRG methods SRG_{SL}, SRG_{ST} and SRG_C yields very good results and performs quite well in all the test-cases. Therefore, we used this combination and the optimal parameters in Table 2 to generate the qualitative results.

Figure 8 shows the results of **Experiment A–D**. First, we show the original gray-scale and color true-orthophoto (gray/RGB), then an overlay with our segmentation (SEG) and finally another overlay with the GT segmentation including

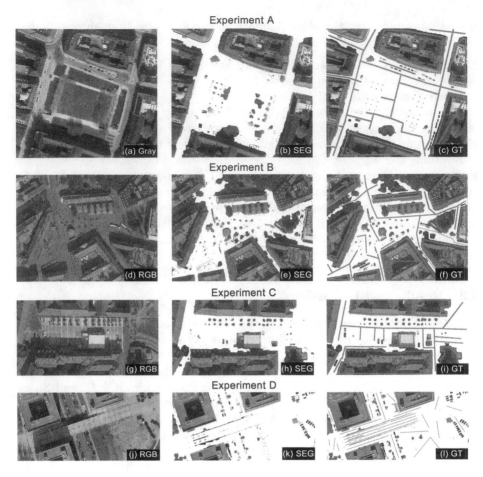

Fig. 8. Experiment A, Marienhof (Munich, Germany), Experiment B, Jakominiplatz (Graz, Austria), Experiment C, Karmeliterplatz (Graz, Austria), Experiment D, Bridge (Hannover, Germany). First column: true-orthophoto; second column: overlay with our segmentation (SEG); third column: overlay with the GT (including the G_P and G_B). (Color figure online)

the annotations of potential evacuation routes GT_P (green) and blocking structures GT_B (red).

With our automated evacuation map extraction algorithm we achieve very good segmentation scores of over 90% in all the real-world experiments. Compared to a manually annotated GT our approach segments a similar area and almost all potential evacuation routes.

In **Experiment A** we get promising results, despite the gray-scale true-orthophoto. All the main exits from the square were segmented correctly, while three narrow streets were left out. During DSM generation the estimation of the height of narrow streets is difficult because their ground levels are only visible

Table 1. The segmentation results of the four experiments with different SRG methods (best scores are depicted in bold). The Dice coefficient (DC), the Jaccard similarity (JS), the segmentation score for potential evacuation routes S_P. Only results where no blocking structures were segmented are shown ($S_B = 100\%$).

SRG	Experiment A			Experiment B			Experiment C			Experiment D		
[%]	DC	JS	S_P	DC	JS	S_P	DC	JS	S_P	DC	JS	S_P
SL	95.74	91.83	94.72	90.52	82.69	93.28	95.70	91.76	92.22	97.51	95.14	99.88
ST	92.40	85.87	93.30	83.53	71.72	87.96	90.70	82.99	91.19	96.26	92.79	97.15
C	1.68	0.85	2.21	29.17	17.07	20.41	18.78	10.36	13.54	1.28	0.64	0.00
SL,ST	95.84	92.01	94.72	90.67	82.93	93.29	95.70	91.76	92.22	**98.24**	**96.54**	99.90
SL,C	95.13	90.72	94.56	**91.83**	84.89	96.02	**97.88**	**95.85**	**99.38**	98.10	96.27	**99.93**
ST,C	95.26	90.95	94.20	91.61	84.52	**96.20**	95.78	91.89	98.76	97.17	94.50	98.16
SL,ST,C	**96.04**	**92.39**	**94.92**	91.83	**84.90**	96.02	**97.88**	**95.85**	**99.38**	98.07	96.21	99.88

Table 2. Optimal parameter settings for T_{stair}, T_{slope}, λ and the spatial resolution $s_{x,y}$ for **Experiments A–D**.

	T_{stair} [cm]	T_{slope} [cm]	λ	$s_{x,y}$ [cm]
Exp. A	20	3.5	4	10
Exp. B	10	3.5	24	10
Exp. C	10	3.0	78	10
Exp. D	10	1.0	24	3

in the few aerial images taken directly above the them. Thus, the surrounding buildings greatly influence the calculation and the ground level appears higher than it actually is.

A similar problem occurs in **Experiment B**, where the street on the top left corner is mostly covered by trees.

Experiment C contains a square with a stage and a tent, which were both correctly left out of the segmentation. An interesting part is the narrow street at the left bottom, where the DSM is not very good, but using the SRG_C approach, we manage to overcome this small obstacle and correctly segment most of the area. This is also the reason for the difference in the score S_P between the combinations involving SRG_C and the ones without.

The spatial resolution in **Experiment D** is by far the best and we are even able to leave most of the cars out of the segmentation, which is quite difficult with a higher resolution of around 10 cm (like in **Experiment A**). The limiting factors in the real-world scenarios are the quality and spatial resolution of the DSMs and true-orthophotos because we calculate the segmentation from these inputs, meaning our results are only as good as the input data.

The choice of T_{stair}, λ **and** T_{slope}.
Our segmentation results depend not only on the quality of the input data, but also on the right parameter settings. Table 2 shows different choices for T_{stair}, λ and T_{slope} for **Experiment A–D** (optimal settings in bold). In our

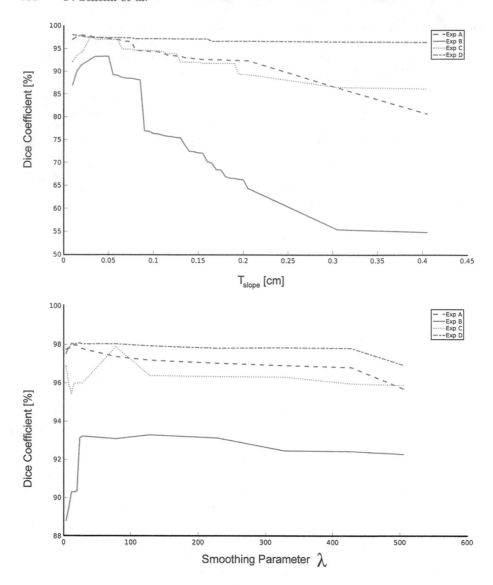

Fig. 9. Two plots showing the DC [%] for different values of T_{slope} (top) and λ (bottom). Especially the choice of T_{slope} is very important for the DC.

experiments, we found that the value of T_{stair} is not critical, thus we focus only on the choices for λ and T_{slope}.

Figure 9 depicts the evaluation of the DC [%] for **Experiment A–D** for $T_{slope} = [0.005, 0.5]$ cm (top) and $\lambda = [4, 500]$ (bottom). The step-sizes are very small in the beginning and widen when the values get higher. For the evaluation of λ we set T_{slope} to the optimal value calculated in Sect. 4.3 and vice versa.

T_{slope} is in general limited by the accessibility of human and a realistic choice is usually around an inclination of 35%, which would mean $T_{slope} = 3.5$ for a spatial resolution of $s_x = s_y = 10$ cm. Higher values can easily lead to an over-segmentation and the WA would include inaccessible areas in that case.

A very small λ gives more emphasize to the smoothing term and we even smooth over edges, which can lead to an over-segmentation, while a higher λ reduces the smoothing effect. We found that values $10 < \lambda < 150$ are in general good choices (see Fig. 4).

5   Conclusion and Outlook

We presented a novel, efficient and easy-to-use approach to extract the WA of outdoor event sites for the purpose of large-scale evacuation simulation. By incorporating information from DSMs, true-orthophotos and inclination maps computed from high-resolution aerial images, we managed to get a very accurate segmentation of the WA area. Further, we showed in the evaluation that a combination of different SRG methods indeed gives better accuracy than a single method. Most common evacuation simulation tools use CAD models, which can be easily created from our WA segmentation. To assess the quality of the generated evacuation maps, we additionally designed a new annotation and scoring scheme.

Despite the very promising results of our approach, there is still room for improvement. One very interesting direction would be to introduce semantic information provided either by a GIS system or some kind of computer vision method to enhance segmentation accuracy for narrow streets. A potential next step is an evaluation by the example of an actual event.

Acknowledgements. This work was financed by the KIRAS program (no 840858, AIRPLAN) under supervision of the Austrian Research Promotion Agency (FFG) and in cooperation with the Austrian Ministry for Traffic, Innovation and Technology (BMVIT).

A   Appendix

In the Appendix, we show a typical evacuation simulation scenario where our generated digital map can be used. We use the map of the Marienhof (Munich, Germany) generated in **Experiment A** and utilize the software tool PedGo [8]. The package comprises three different programs:

- PedEd - Used for editing the map, placing persons and marking the exits
- PedGo - The simulation program, where various scenarios can be simulated
- PedView - A 3D visualization of the previously calculated simulations

The first step is always loading the map into the editor PedEd and placing the exits (see Fig. 10, left). They are usually at the end of the streets leading away from the central area. After that, persons (or agents) can be put onto the map and corrections to the map can be made. The whole process usually takes less than three minutes. The next step is starting the simulation tool (PedGo) and loading the project. To get an estimate of the average evacuation time, multiple simulations should be performed (see Fig. 10, right). With PedView we can then view simulation files generated with PedGo in full 3D (see Fig. 11).

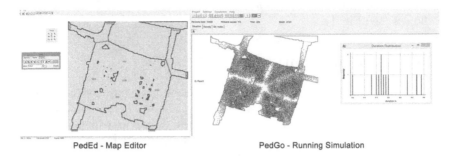

PedEd - Map Editor   **PedGo - Running Simulation**

Fig. 10. With PedEd the extracted CAD model can be edited and then various simulations can be performed with PedGo.

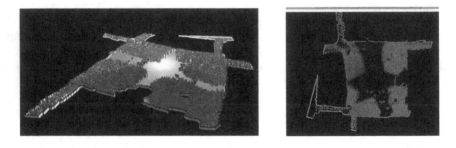

Fig. 11. PedView can present the simulations calculated with PedGo in 3D.

References

1. Helbing, D., Mukerji, P.: Crowd disasters as systemic failures: analysis of the love parade disaster. EPJ Data Sci. **1**, 1–40 (2012)
2. Krausz, B., Bauckhage, C.: Loveparade 2010: automatic video analysis of a crowd disaster. Comput. Vis. Image Underst. **116**, 307–319 (2012)
3. Hsu, E.B., Burkle, F.M.: Cambodian bon om touk stampede highlights preventable tragedy. Prehospital Disaster Med. **27**, 481–482 (2012)
4. Taubenböck, H., Post, J., Kiefl, R., Roth, A., Ismail, F.A., Strunz, G., Dech, S.: Risk and vulnerability assessment to tsunami hazard using very high resolution satellite data: the case study of padang, indonesia. EARSeL eProceeding **8**, 53–63 (2009)

5. Lämmel, G., Grether, D., Nagel, K.: The representation and implementation of time-dependent inundation in large-scale microscopic evacuation simulations. Transp. Res. Part C Emerg. Technol. **18**, 84–98 (2010)
6. Schenk, F., Rüther, M., Bischof, H.: Automated segmentation of the walkable area from aerial images for evacuation simulation. In: Proceedings of the 2nd International Conference on Geographical Information Systems Theory, Applications and Management, pp. 125–135 (2016)
7. Galea, E.R.: Simulating evacuation and circulation in planes, trains, buildings and ships using the exodus software. In: Pedestrian and Evacuation Dynamics, pp. 203–225. Springer (2002)
8. Klüpfel, H.: The simulation of crowds at very large events. In: Schadschneider A., Pöschel, T., Kühne, R., Schreckenberg, M., Wolf, D.E. (eds) Traffic and Granular Flow 2005. Springer, Heidelberg (2007)
9. Tsai, J., Fridman, N., Bowring, E., Brown, M., Epstein, S., Kaminka, G., Marsella, S., Ogden, A., Rika, I., Sheel, A., et al.: Escapes: evacuation simulation with children, authorities, parents, emotions, and social comparison. In: AAMAS, pp. 457–464 (2011)
10. Schneider, V., Könnecke, R.: Simulating evacuation processes with aseri. In: Pedestrian and Evacuation Dynamics, pp. 301–313 (2001)
11. Johnson, C.W.: Olympic venues. Safety Sci. **46**(2008), 302–322 (2012)
12. Shi, C., Zhong, M., Nong, X., He, L., Shi, J., Feng, G.: Modeling and safety strategy of passenger evacuation in a metro station in China. Safety Sci. **50**, 1319–1332 (2012)
13. Tang, F., Ren, A.: GIS-based 3d evacuation simulation for indoor fire. Build. Environ. **49**, 193–202 (2012)
14. Mas, E., Adriano, B., Koshimura, S.: An integrated simulation of tsunami hazard and human evacuation in la punta, peru. J. Disaster Res. **8**, 285–295 (2013)
15. Koch, D.B., Payne, P.W.: An incident management preparedness and coordination toolkit. In: Global Humanitarian Technology Conference (GHTC), pp. 31–35. IEEE (2012)
16. Dal Poz, A.P., Gallis, R.A., da Silva, J.F., Martins, É.F.: Object-space road extraction in rural areas using stereoscopic aerial images. IEEE Geosci. Remote Sens. Lett. **9**, 654–658 (2012)
17. Hu, J., Razdan, A., Femiani, J.C., Cui, M., Wonka, P.: Road network extraction and intersection detection from aerial images by tracking road footprints. IEEE Trans. Geosci. Remote Sens. **45**, 4144–4157 (2007)
18. Lin, Y., Saripalli, S.: Road detection and tracking from aerial desert imagery. J. Intell. Robot. Syst. **65**, 345–359 (2012)
19. Zhou, H., Kong, H., Wei, L., Creighton, D., Nahavandi, S.: Efficient road detection and tracking for unmanned aerial vehicle. IEEE Trans. Intell. Transp. Syst. **16**, 297–309 (2015)
20. Cheriyadat, A.M.: Unsupervised feature learning for aerial scene classification. IEEE Trans. Geosci. Remote Sens. **52**, 439–451 (2014)
21. Huth, J., Kuenzer, C., Wehrmann, T., Gebhardt, S., Tuan, V.Q., Dech, S.: Land cover and land use classification with twopac: towards automated processing for pixel-and object-based image classification. Remote Sens. **4**, 2530–2553 (2012)
22. Han, J., Zhang, D., Cheng, G., Guo, L., Ren, J.: Object detection in optical remote sensing images based on weakly supervised learning and high-level feature learning. Trans. Geosci. Remote Sens. **53**, 3325–3337 (2015)
23. Adams, R., Bischof, L.: Seeded region growing. IEEE Trans. Pattern Anal. Mach. Intell. **16**, 641–647 (1994)

24. Irschara, A., Rumpler, M., Meixner, P., Pock, T., Bischof, H.: Efficient and globally optimal multi view dense matching for aerial images. ISPRS Ann. Photogrammetry Remote Sens. Spatial Inf. Sci. **1**, 227–232 (2012)
25. Lowe, D.G.: Distinctive image features from scale-invariant keypoints. Int. J. Comput. Vis. **60**, 91–110 (2004)
26. Fischler, M.A., Bolles, R.C.: Random sample consensus: a paradigm for model fitting with applications to image analysis and automated cartography. Commun. ACM **24**, 381–395 (1981)
27. Triggs, B., McLauchlan, P.F., Hartley, R.I., Fitzgibbon, A.W.: Bundle adjustment — a modern synthesis. In: Triggs, B., Zisserman, A., Szeliski, R. (eds.) IWVA 1999. LNCS, vol. 1883, pp. 298–372. Springer, Heidelberg (2000). doi:10.1007/3-540-44480-7_21
28. Rumpler, M., Wendel, A., Bischof, H.: Probabilistic range image integration for DSM and true-orthophoto generation. In: Kämäräinen, J.-K., Koskela, M. (eds.) SCIA 2013. LNCS, vol. 7944, pp. 533–544. Springer, Heidelberg (2013). doi:10.1007/978-3-642-38886-6_50
29. Collins, R.T.: A space-sweep approach to true multi-image matching. In: Computer Vision and Pattern Recognition, pp. 358–363. IEEE (1996)
30. Holz, D., Holzer, S., Rusu, R.B., Behnke, S.: Real-time plane segmentation using RGB-D cameras. In: Röfer, T., Mayer, N.M., Savage, J., Saranlı, U. (eds.) RoboCup 2011. LNCS, vol. 7416, pp. 306–317. Springer, Heidelberg (2012). doi:10.1007/978-3-642-32060-6_26
31. Rudin, L., Osher, S., Fatemi, E.: Nonlinear total variation based noise removal algorithms. Physica D Nonlinear Phenom. **60**, 259–268 (1992)
32. Chambolle, A., Pock, T.: A first-order primal-dual algorithm for convex problems with applications to imaging. J. Math. Imag. Vis. **40**, 120–145 (2011)
33. Dice, L.R.: Measures of the amount of ecologic association between species. Ecology **26**, 297–302 (1945)
34. Jaccard, P.: Nouvelles recherches sur la distribution florale (1908)

Information System for Automated Multicriterial Analytical Control of Geomagnetic Field and Space Weather Parameters

Andrei Vorobev[✉] and Gulnara Vorobeva (Shakirova)

Ufa State Aviation Technical University, Ufa, Russia
geomagnet@list.ru

Abstract. Long-term experimental researches in various spheres prove an influence of space weather, geomagnetic field, its variations and anomalies on systems and objects of various origins. The estimation of the influence requires an effective approach to analyze the principles of distribution of geomagnetic field parameters on the Earth's surface, its subsoil and in circumterrestrial space. In this paper the authors suggest the solution, which is based on modern geoinformation and web technologies and provides the mechanisms to calculate, analyze and visualize parameters of geomagnetic field and its variations. The most attention here is paid to research and development of integrated information space for monitoring and analytical control of parameters of space weather, geomagnetic field and its variations. The integrated information space is based on the set of conception, models, and methods and represented as analytical information system GEOMAGNET (https://www.geomagnet.ru).

Keywords: Geoinformation systems · Geomagnetic variations · Analytical information systems · Geomagnetic field · Space weather · Magnetic storms · Data assimilation · Integration and fusion

1 Introduction

Long-term experimental researches in various spheres prove an influence of some single geomagnetic variations (GMV) components or their combination on biological, technical, geological and other objects and systems. The influence can be both direct or through some agents. As a result the system is forced to either adapt to magnetic state changes (with deformation, mutation and so on) or exist in stress, unstable or other abnormal mode.

Today geomagnetic field (GMF) and its variations parameters are partially studied and monitored by magnetic observatories. A magnetic observatory is a special scientific organization, where parametric and astronomical observations of the Earth's magnetosphere are performed.

According to NOAA (National Oceanic and Atmospheric Administration), BGS (British Geological Survey), Schmidt Institute of Physics of the Earth of the Russian Academy of Sciences (IPE RAS), Pushkov Institute of Terrestrial Magnetism,

C. Grueau et al. (Eds.): GISTAM 2016, CCIS 741, pp. 109–121, 2017.
DOI: 10.1007/978-3-319-62618-5_7

Ionosphere and Radio Wave Propagation of the Russian Academy of Sciences (IZMIRAN) the majority of magnetic observatories is concentrated in Europe.

Another problem is a lack of free-access effective technologies and ergonomic tools to provide geomagnetic and space weather data integration into integrated information space.

The registered information is regularly sent to the International centers in Russia, USA, Denmark and Japan, where the information is registered, analyzed and partially available to the broader audience with some delay. All data measured and collected about geomagnetic field is distributed in various sources and storages. As a result there is no easy-to-use and effective tool for operative monitoring, analytical control, modeling and visualization of geomagnetic and space weather data for specialists in various spheres.

The obvious way to solve the problem is to implement innovative information technologies there. In particular the most expectations are about using geoinformation systems and technologies to solve the problem. In this paper the authors suggest an approach to study, monitoring, analyze and visualize space weather, geomagnetic field, its variations and anomalies, which is based on modern Web and geoinformation technologies.

2   Brief Overview of Space Weather, Geomagnetic Field and Its Variations

Space Weather. The first official definition of the term "space weather" was given in 1995 in the United States during development of National Space Weather Program (NSWP). According to the definition, "space weather" is a set of changes on the Sun, in Solar Wind, magnetosphere and ionosphere, which can influence on board and ground technical systems efficiency and reliability as well as to harm human life and health. The most common registered parameters of the solar flux are velocity, temperature and concentration of the solar wind (GOST 25645.136-86).

In recent years a new term "space climate" refers to the long-period variations of solar activity and space weather. Due to the dynamic development of space programs an analytical control of space climate parameters becomes increasingly important.

Geomagnetic Field and Its Variations. GMF structure is complicated and inhomogeneous, that is why its distribution on the Earth's surface has an anisotropic character. So, induction of GMF on the border of the magnetosphere is ~ 10.03 mT. It is 20–30 mT at the equator and 60–70 µT at the poles near the Earth's surface (Fig. 1) [1].

Here the main field total intensity vector in the point with given spatiotemporal coordinates (latitude, longitude, elevation, year) is defined as a sum of three components:

$$B_{\text{GE}} = B_{\text{int}} + B_{\text{ext}} + B_{\text{tech}},$$

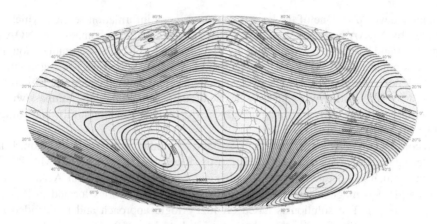

Fig. 1. Main field total intensity distribution.

where B_{int} is a vector of GMF of intraterrestrial sources; B_{ext} is a variable component of GMF, which is caused by influence of external factors and geomagnetic disturbances; B_{tech} is a GMF intensity vector component with technogenic (anthropogenic) origin.

Magnetic field of intraterrestrial sources consists of two fields: electric currents field in the core (the main ("core") field), which is $\sim 98\%$ of the whole field, and rocks magnetism field, which is $\sim 2\%$ of the whole field. Thus, with altitude raise the crust field decreases much quicker, then the main field, and from the altitude ~ 100 km its value can be neglected.

The calculated or some averaged values of GMF parameters are conditionally accepted as a normal (undisturbed) state of geomagnetosphere. Deviations of actual values of GMF parameters from these normal values are known as geomagnetic variations (GMV) and are estimated by special geomagnetic indices [10, 11].

3 Overview and Analysis of the Problem Solutions

Information technologies (IT), which provide methods and tools for operating geospatial data, are among the most dynamic technologies. This popularity has two reasons: firstly, 80% of the data has a geospatial reference; secondly, it correlates with fast grow of SQL- and noSQL-technologies.

Parameters of space weather, GMF and its variations have a geospatial reference and can be classified as spatial data. So, this data can be operated by the appropriate information technologies, which provide geospatial reference of attributive data, their processing, analysis and visualization.

In spite of wide variety of software solutions and technologies a solution of the problem of digital processing and visualization of parameters of space weather, GMF and GMV is still not or just partially supported by IT and GIS. At the same time an efficiency of this approach to the problem solution is obvious.

The analysis of the problem solutions proved, that information technologies providing calculation, geospatial reference, visualization and automated data processing are poorly developed. One of the known solutions is the information service, which is provided by NOAA and available at http://www.ngdc.noaa.gov/geomag-web [NOAA Magnetic Field Calculators]. Although the calculation results are acceptable enough, the solution has a lot of disadvantages, such as no visualization tools, low ergonomics and efficiency. The same disadvantages are inherent to other similar services, which are provided by BGS (http://www.geomag.bgs.ac.uk/data_service/models_compass/wmm _calc.html), CIRES (http://geomag.org/models/igrfplus-declination.html) and other leading organizations.

According to geomagnetic data analysis there is another well-known solution. It is the Interregional Geomagnetic Data Center of the Russian-Ukrainian INTERMAGNET segment, which is operated by the Geophysical Center of the Russian Academy of Sciences [4]. Geomagnetic data are transmitted from observatories located in Russia and Ukraine [4]. The solution is based on fuzzy logic approach and is intended to real-time recognition of artificial (anthropogenic) disturbances in incoming data.

One more key element of the mentioned problem is concerned with a number of heterogeneous data sources, which collect and share historical and current values of parameters of space weather, geomagnetic field and its variations. These are powerful web services provided by INTERMAGNET [3], BGS [2] and other organizations. The main thing here is that all these data sources are not integrated. There are no any tools, which provide to user a single access to space weather and geomagnetic data.

In this paper the authors suggest a solution of the problem – the analytical information system (AIS) GEOMAGNET, which is available at https://www.geomagnet.ru. AIS GEOMAGNET provides modeling, monitoring, analytical control, two- and three-dimensional visualizations of space weather, GMF and its variations parameters. With this system a user can get access to data, automatically analyze them and use the results for solving applied problems [6–9].

4   Conception and Tools of AIS GEOMAGNET

Figure 2 represents a conception of the system of monitoring and control of parameters of space weather, GMF and its variations. The conception is based on a paradigm of heterogeneous data sources combination into an integrated information space with platform-independent mechanism of their analytical control and interpretation.

AIS GEOMAGNET provides the main mechanisms of real-time monitoring, modeling, analytical control and forecast of space weather, GMF and its variations parameters (discretization step is 1 min). It is based on relevant available data and primarily provides an improvement of informativeness, efficiency, and ergonomics criteria in monitoring and analytical control of space weather, GMF and GMV parameters.

GEOMAGNET provides and is based on the following GIS mechanisms:

- direct and reverse geocoding of spatial data;
- interpretation and visualization of spatial data;
- geolocation (the identification of exact users device location);

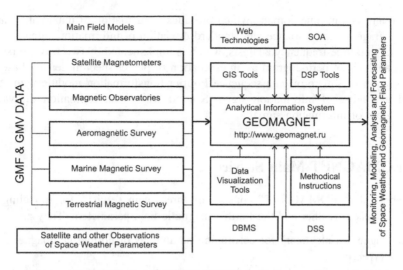

Fig. 2. Conception of AIS GEOMAGNET (https://www.geomagnet.ru).

- processing, visualization, translation and analysis of available magnetic observatories data (with support of INTERMAGNET and Geophysical Center of Russian Academy of Sciences);
- virtual globe based on WebGL technology.

AIS GEOMAGNET also supports the following mechanisms of data visualization and interpretation:

- multilayer visualization of spatial data with optional support of user-defined KML/KMZ layers [Patent for Invention No 2568274, 2015];
- visualization of two-dimensional array with dynamic time axis scalability on the basis of web interfaces D3.js – Data-Driven Documents;
- forecast (detailed for 3 days and daily-average for 27 days) of space weather and GMF parameters;
- 2D и 3D-visualization of geospatial data with scale-depend detalization level:

$$n = \log_2(\log_4 m) + 2,$$

where m is a quantity of tiles; n is a discrete scale level.

AIS GEOMAGNET uses classic mechanisms of digital signal processing (DSP) to provide a set of tools for digital processing of data about space weather, GMF and GMV parameters, such as following:

- linear, nonlinear and adaptive filtering of information signal, which provides noise reduction and signal bandpassing in frequency domain as well as analysis of correlation of adjacent information signals parameters;

- signal analysis in time domain, which provides a calculation of maximal, minimal and average values, variance and standard deviation of information signal;
- spectral and time-frequency analysis, which is implemented by two ways. First way is an analysis of periodogram as an estimation of power spectral density based on calculation of square of data sequence Fourier transformation module with statistical averaging of information signa. Second way is an estimation of wavelet scalogram of information signal.

5   GEOMAGNET Main Services

The webportal "GEOMAGNET" consists of three main services for modeling and analysis of space weather and the Earth's magnetic field parameters. They are

- "Geomagnetic Calculator" for calculating parameters of normal geomagnetic field in the given point,
- "Solar Activity" for analyzing 3-days space weather parameters,
- "Magnetic Stations" for integrating data registered by magnetic stations all around the world.

5.1   "Geomagnetic Calculator (WMM2015)" Service

Geomagnetic (or magnetic) calculator is a special type of a web application, which provides an estimation of magnetic field at a given point on Earth.

The "Geomagnetic Calculator (WMM2015)" is a web-based application within "GEOMAGNET" WebGIS application providing calculation parameters of geomagnetic field and its secular variations according to the set of coordinates and dates (Fig. 3). A user of any level can calculate and analyze parameters of geomagnetic field at his current location or at any other point on the Earth [8].

The "Geomagnetic Calculator (WMM2015)" is a responsive application, which does not depend on device type and parameters but is transferable and works similarly in any device, i.e. phones, tablets, desktops. This platform- and device-independency is

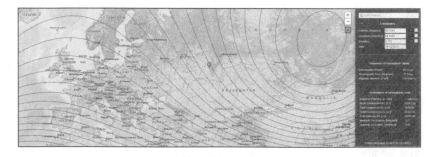

Fig. 3. Interface of the "Geomagnetic Calculator (WMM2015)".

realized on the basis of the special framework "Bootstrap" (http://getbootstrap.com) that is a set of programming libraries for HTML, CSS and JavaScript. Flexibility and performance of this application tool are also increased by supporting HTML5 and CSS3 standards.

The main functionality of the "Geomagnetic Calculator (WMM2015)" provides effective and reliable calculation and analysis of parameters of normal (undisturbed) geomagnetic field by the spatiotemporal coordinates with an error value of no more than 0.1%.

"Geomagnetic Calculator (WMM2015)" provides the calculation of the following parameters of geomagnetic field:

- north component of geomagnetic field induction vector;
- vertical component of geomagnetic field induction vector;
- magnetic declination and dip;
- scalar potential of geomagnetic field induction vector.

5.2 "Solar Activity" Service

The web application "Solar Activity: Monitoring and Analysis" (https://www.geomagnet.ru/solar_activity.html) allows users to monitor solar weather in real time on the basis of satellite observations and carry out a comprehensive analysis of its parameters, to assess how their values change for different periods of time (from one hour to several days). The service takes data via asynchronous requests to the web server and renders results in real time mode both in textual and graphic forms. Data analysis is also available in amplitude and frequency mode by single or groups of parameters of space weather.

With "Solar Activity" service a user gets last 3-days information about space weather parameters. In particular there are three main available data parameters:

- Proton density $[cm^{-3}]$;
- Temperature of Solar Wind $[° K]$;
- Velocity of Solar Wind [km/sec].

The main window of the service is divided into two main parts – statistical and graphical.

In the statistical part the results of signal analysis in time domain are represented. It contains such values as maximal, minimal and average values, variance and standard deviation, which are calculated for proton density, temperature and velocity of the solar wind. Also here the current values of these parameters are displayed (Fig. 4).

Also in this part of "Solar Activity" window there is a characteristic of magnetic state according to the Kp-index scale. According to this scale there are five possible magnetic states:

- minor magnetic storm;
- moderate magnetic storm;
- strong magnetic storm;
- severe magnetic storm;
- extreme magnetic storm.

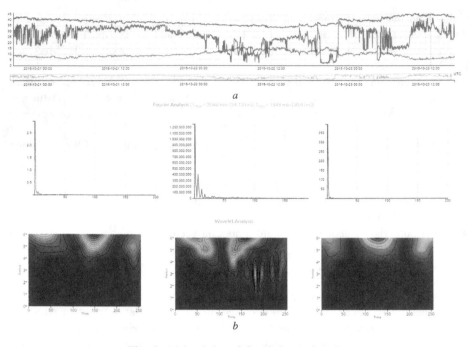

Fig. 4. Main charts of the "Solar Activity".

These values describe how current and future space weather conditions can influence on systems of various origins, especially on human. Each value has a corresponding color: from green to deep red. The same color is used in the "Solar Activity" interface for indicating current geomagnetic state.

Another part of the "Solar Activity" window is a graphical one. It provides a set of charts, which display the results of linear and nonlinear filtering of space weather data in time and frequency domain as well as spectral and time-frequency analysis.

5.3 "Magnetic Stations" Service

Today magnetic stations connected to integrated international network are the main way to observe geomagnetic field. With period of a few hours all data registered by stations is sent to World centers of collecting and processing of geomagnetic data:

- World Data Center for Solar-Terrestrial Physics, Moscow;
- International Space Weather Center, which is regionally distributed in Brussels, Boulder, Moscow, Tokyo, Sydney and Beijing;
- etc.

Geomagnetic data is represented in various formats, stored in a number of data sources and it is hard to make queries to get them.

To join them all together the authors suggested another service in web-portal "GEOMAGNET", which provides an integrated access to all geomagnetic data from all available magnetic observatories. This service is called "Magnetic Stations" (https:// www.geomagnet.ru/magnetic_stations.html).

The service has a single user interface, which provides a map displaying the locations of all available magnetic observatories. Due to this type of observatories representation a user can choose any magnetic station in any location without a necessity of knowing its name or exact coordinates (Fig. 5). Also a user can pick up an observatory and get its registered data.

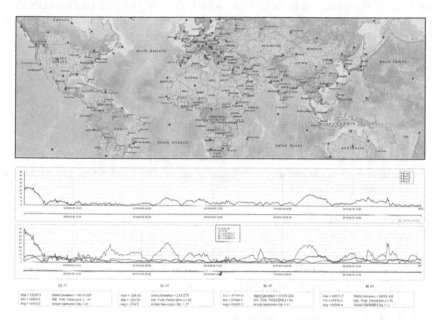

Fig. 5. Magnetic observatories on the map and their analysis result in the "Magnetic Stations".

After page loading a user gets information about an amount of both active and inactive magnetic stations. All active observatories are colored red and inactive ones are colored grey to separate them from each other.

General information about the magnetic station (name, country, geographic coordinates, etc.) is available in popup windows, which is displayed near the observatory icon when a user moves a mouse pointer over it. This information is available for both active and inactive magnetic observatories.

More detailed information about a magnetic station supposes visualization and analysis of data registered by the observatory for recent 3 days. The link to the separate page with data analysis for the observatory is available through the popup window with general information (the link is displayed only if the magnetic station is active; for inactive observatories only general information is available).

For the user chosen magnetic observatory the following parameters of geomagnetic field are visualized and analyzed within the service:

- north component (X) [nT];
- east component (Y) [nT];
- down component (Z) [nT];
- total intensity [nT].

6   Data Analysis in GEOMAGNET

Tables 1 and 2 represent methods of DSP, which are used in AIS GEOMAGNET.

Table 1. Main methodic of DSP, which is used in AIS GEOMAGNET.

#	Step
1	Reading the discrete information signal (IS) as JSON/CSV/XML array
2	Detection, analysis and exclusion of missing measurements
3	Calculation and exclusion of IS constant component
4	Calculation of IS variance
5	Calculation of numerical value of IS frequency threshold detector
6	Calculation of direct Discrete Fourier Transformation (DFT) and periodogram based on Fast Fourier Transform (FFT)
7	Calculation of the biggest IS frequency with power spectral density less than IS frequency threshold detector
8	Calculation of low frequencies digital filter cutoff frequency
9	Preparation of discrete IS (original array) for digital filtering
10	Calculation of direct DFT of IS, which is prepared to filtering
11	Multiplication of Fourier image and transfer function
12	Calculation of reverse DFT of IS on the basis of FFT
13	Preparation of the array to save/output
14	Save/output of the filtered array in JSON/CSV/XML format

It is necessary to mention, that calculation and exclusion of the constant component on the step 3 is performed according to the following expression:

$$x_k^* = x_k - \frac{1}{N}\sum_{k=0}^{N-1} x_k, \tag{1}$$

where x_k^* is k-th element of array with constant component excluded from source array x_k; N is a length of array.

Next standard deviation and variance of information signal are calculated according to the following expression:

$$\sigma^2 = \frac{1}{N-1}\sum_{k=0}^{N-1} \left(x_k^*\right)^2, \tag{2}$$

Table 2. Methodic of information signal time-frequency analysis.

#	Step name
1	Reading of discrete information signal (IS) as array
2	Calculation of IS minimal and maximum values
3	Calculation and exclusion of IS constant component
4	Calculation of IS standard deviation and variance
5	Calculation of IS Fourier transformation and periodogram
6	Calculation of IS frequency threshold detector
7	Calculation of parameters of discretization of IS discrete wavelet transformation
8	Calculation of IS discrete wavelet transformation and scalogram
9	Graphical interpretation of IS discrete wavelet transformation and scalogram

where σ is a standard deviation and σ^2 is a variance of information signal.

On step 5 (Tables 1 and 2) a numerical value of information signal frequency threshold detector l is calculated according expression (3):

$$l = \frac{\sigma^2 X_q}{N}, \text{ where } X_q = -\ln(q). \tag{3}$$

This value is an indicator that the harmonic exists in information signal not as a component of white noise. Probability of such a statement is $1 - q$, where $q \ll 1$ (for example, in this case $q = 0.03$).

Fourier transformation of discrete function with excluded constant component is used on step 6 (Tables 1 and 2) to calculate an information signal periodogram with expression (4). In case of odd N the upper limit of calculation is $(N + 1)/2$:

$$D_j = \frac{1}{N^2}\left[\left(\operatorname{Re} X_j\right)^2 + \left(\operatorname{Im} X_j\right)^2\right], j = 0, 1, \ldots, N/2, \tag{4}$$

where D_j is j-th element of periodogram; $\operatorname{Re} X_j$ and $\operatorname{Im} X_j$ are real and imaginary parts of Fourier transformation, which are calculated as follows:

$$X_j = \sum_{k=0}^{N-1} x_k^* e^{-i\frac{2\pi}{N}kj}, j = 0, 1, \ldots, N - 1,$$

So, if Δt is a discretization step of source signal, then periodogram counts refer to the frequencies:

$$v_j = \Delta v \cdot j, j = 0, 1, \ldots, N/2, \text{ where } \Delta v = \frac{1}{N\Delta t}.$$

Next, on step 7 (Table 2) the discretization parameters for discrete Fourier transformation are calculated. So, for example, rounding up value of maximal scale level $a_{\max}$ is calculated as follows:

$$a_{\max} = \log_2\left(\frac{N}{2}\right).$$ (5)

Next (Table 2) a discrete wavelet transformation of information signal is calculated according to the expression (6):

$$W(k, a_i, b_j) = \sum_{-N/2}^{N/2} \left[\frac{1}{\sqrt{a_i}} \cdot x_k^* \cdot \Psi\left(\frac{k - b_j}{a_i}\right)\right], \text{where } a_i = a_0^m; \ a_0 = 2; \ m$$
$$= 0, 1, \ldots, a_{\max};$$ (6)

As well as its scalogram:

$$S(a_i, b_j) = \left|W(a_i, b_j)^2\right|;$$ (7)

where Ψ is a parent wavelet; a is a scale parameter; b is a shift parameter ($b = -N/2$, ..., $N/2$).

Steps 9 и 13 (Table 1) are concerned with preparations of information signal to digital filtering and a source array to save/output. In both cases the preparation is to multiply each element of the array by value $(-1)^i$, where i is an array element index. This operation provides harmonics count from the origin of coordinate system.

AIS GEOMAGNET uses for data interpretation API D3.js (Figs. 4 and 5). It is a JavaScript library, which provides data analysis and visualization for Web. Because of its platform-independence and web-focus AIS GEOMAGNET is available for the users with any platform and device.

7 Conclusions

During the research the authors have suggested and developed a conception of modeling and distant analytical control of space weather, GMF and GMV parameters. The conception is based on a paradigm of heterogeneous data sources (satellite, air, marine, ground observations, and magnetic observatories) combination into integrated information space with platform-independent mechanisms of DSP and data interpretation. The research proved, that a program formalization of the conception as AIS GEOMAGNET reduces the time required for collecting, processing and analyzing heterogeneous data about space weather, GMF and GMV.

Acknowledgements. The reported study was supported by RFBR, research projects No. 14-07-00260-a, 14-07-31344-mol-a, 15-17-20002-d_s, 15-07-02731_a, and the grant of President of Russian Federation for the young scientists support MK-5340.2015.9.

References

1. Aminatov, A.S., et al.: Variations of the Earth's magnetic field, 52 p. StroiArt, Moscow (2001). (in Russian)
2. Dawson, E., Lowndes, J., Reddy, P.: The British geological survey's new geomagnetic data web service. Data Sci. J. **12**, WDS75–WDS80 (2013)
3. Kerridge, D.: INTERMAGNET: worldwide near-real-time geomagnetic observatory data. http://www.intermagnet.org/publications/IM_ESTEC.pdf
4. Soloviev, A., Bogoutdinov, S., Gvishiani, A., Kulchinskiy, R., Zlotnicki, J.: Mathematical tools for geomagnetic data monitoring and the INTERMAGNET Russian segment. Data Sci. J. **12**, WDS114–WDS119 (2013). doi:10.2481/dsj.WDS-019
5. Thomson, A.: Geomagnetic observatories: monitoring the Earth's magnetic and space weather environment. Weather **69**(9), 234–237 (2014)
6. Vorobev, A.V., Shakirova, G.R.: Applicaton of geobrowsers to 2D/3D-visualisation of geomagnetic field. In: Proceedings of the 15th SGEM GeoConference on Informatics, Geoinformatics and Remote Sensing, Albena, 18–24 June 2015, vol. 1, pp. 479–486 (2015)
7. Vorobev, A.V., Shakirova, G.R.: Modeling and 2D/3D-visualization of geomagnetic field and its variations parameters. In: Proceedings of GISTAM 2015 - 1st International Conference on Geographical Information Systems Theory, Applications and Management, pp. 35–42 (2015)
8. Vorobev, A.V., Shakirova, G.R.: Web-based information system for modeling and analysis of parameters of geomagnetic field. Procedia Comput. Sci. **59**, 73–82 (2015)
9. Vorobev, A.V., Shakirova, G.R.: Calculation and analysis of dynamics of main magnetic field parameters for the period 2010–2015. Geoinformatika **1**, 37–46 (2015)
10. Yanovsky, B.M.: The Earth's magnetism, 590 p. Leningrad (1978). (in Russian)
11. Zabolotnaya, N.A.: Geomagnetic activity indices, 84 p. LKI, Moscow (2007). (in Russian)

SWRL Rule Development to Automate Spatial Transactions in Government

Premalatha Varadharajulu[1,2]($\boxtimes$), Lesley Arnold[1,2], David A. McMeekin[1,2], Geoff West[1,2], and Simon Moncrieff[1,2]

[1] Curtin University, Perth, WA 6845, Australia
p.varadharajulu@postgrad.curtin.edu.au
[2] Cooperative Research Centre for Spatial Information, Carlton, Australia

Abstract. The land development approval process between local councils and government planning authorities is time consuming and resource intensive because human decision-making is required to complete a transaction. This is particularly apparent when seeking approval for a new land subdivisions and administrative boundary changes that require changes to spatial datasets. This paper presents a methodology that automates the approval process by developing. Feedback on the transaction is communicated to the land developer in real-time, thus reducing process handling time for both developer and the government agency. This paper presents an approach for knowledge acquisition on rule development using Semantic Web and Artificial Intelligence to automate the spatial transaction process. The Web Ontology Language (OWL) is used to represent relationships between different entities in the spatial database schema. Rules that replicate human knowledge are extracted from government policy documents and subject-matter experts, and are defined in the form of Semantic Web Rule Language (SWRL) and based on geometry and attributes of database entities. The SWRL rules work with OWL-2 (spatial schema and vocabulary) ontologies to enable the automatic transactions to occur. These rules are implemented using an ontology and rule reasoner, which accesses the instances of data elements stored in the underlying spatial database. When the developer submits an application, the software checks the rules against the request for compliance with the relevant government policies and standards. This paper presents results for dealing with road proposals and road name approvals.

Keywords: Spatial transaction · Spatial data supply chain · Artificial intelligence · Semantic Web · Ontology · Rule-based reasoning · OWL-2

1 Introduction

Land developers and local government authorities are required to submit proposals for new subdivisions to land and planning departments for approval. These new subdivisions include new land parcel boundaries, roads and road names, and changes to local authority boundaries. The approval process often spans many

C. Grueau et al. (Eds.): GISTAM 2016, CCIS 741, pp. 122–142, 2017.
DOI: 10.1007/978-3-319-62618-5_8

work teams and new information, such as property addresses may need to be generated. This manual process can be time consuming and resource intensive.

New methods are required to reduce data handling and support the automation of transactions with government. Current workflows are characterised by several decision points and a trail of paper documents are often created to formalise the decision-making process and to provide a reference point for legal transactions further along the land administration process [1]. As a result, there is often a time delay of several weeks during which a new subdivision is considered by authorities from the various land development and planning perspectives.

This research seeks to automate the spatial transaction process using artificial intelligence with ontologies to create rules that replace the human decision-making process for land development approvals. A case study examining new road proposals, road names and land administration boundary changes is used to demonstrate the approach. This research is being conducted in conjunction with the Western Australian Land Information Authority (Landgate). Landgate is the approving authority for all new subdivisions in Western Australia, and is responsible for land administration boundary changes resulting from land development activity.

The Semantic Web was first introduced by Tim Berners-Lee who imagined it as "a web of data that can be processed directly and indirectly by machines" [2]. This research is inspired by the increased bandwidth of the Internet and advances in Semantic Web technologies, which now make it possible to automate many of the human elements of the decision-making process on the Web.

Rule-based systems have been used for decision support in the past but these are typically closed client bases systems. However the advantage of the Semantic Web is that the data, ontologies and rules are described using well defined standards (w3c.org) and can be made available over the Web as published resources, typically in one of a number of machine (and human) readable formats [3]. The vision of the Semantic Web is that, ontologies, especially those of a general nature, can be shared and re-used in many applications. In our case, it is envisaged that once a working solution for the approvals process has been validated for one jurisdiction (Western Australia), the ontologies and rules can be used in other jurisdictions (Victoria, New South Wales etc.) and domains.

The work is part of a research program into Spatial Data Infrastructures being conducted at the Cooperative Research Centre for Spatial Information (CRCSI), Australia. One of the objectives of the research program is to automate spatial data supply chains from end-to-end to enable access to the right data, at the right time, at the right price [4].

This research is focusing on the first stage in the spatial data supply chain process, which is the creation of spatial data generated through a land development business process. Instead of paper-based systems, the method enables the capture of spatial information in machine-readable form at its inception point. This is a significant step towards achieving downstream workflow automation. It also supports the recording of data provenance in machine-readable form at the commencement of a spatial transaction to support legal and data quality attribution.

The development consists of two stages. In the first stage, a GUI-based interactive system called Protégé is used to design ontologies and rules from spatial data schema and various documents including policies. The second stage uses a runtime environment (Jena and Java) to process the ontologies and rules along with existing and proposed road data to determine compliance with policies etc.

2   Background and Related Research

Methods for spatial data processing and integration have been researched and developed over the past few years, however little work has considered the automation of the decision-making process using the semantic web where spatial data is an input to the approval process.

One of the objectives of the Semantic Web is to evolve into a universal medium for information, data and knowledge exchange, rather than just being a source for information. To attain this, it uses the well known http protocol and technologies [5,6], such as URIs (Universal Resource Identifiers), RDF (Resource Description Framework) and ontologies with reasoning and rules.

One of the most important components is the RDF, which is a language for representing information about resources on the Web (http://www.w3.org/RDF/). RDF aims to organize information in a machine-readable format by representing information as triples: <subject, predicate, object>, a concept from the artificial intelligence community.

Traditionally, data is generally stored in relational databases. This has been a suitable model for the last few decades as it enables reasonable computers to store the data and allow searching. The advantage is that each piece of data is only stored in one place and each piece of data is atomic. The disadvantage is that the database tables have to be developed in advance usually from entity relational diagrams, the tables do not naturally relate to reality, and it is hard to link various databases together, especially if they are across different systems.

A more natural representation for the Internet (and Web) is the network or graph model. Data items are defined as nodes and the relationships defined as the arcs. A graph can represent anything and allow different pieces of disparate data to be related to each other. Extra links can be added on the fly without the need to redefine databases. For spatial data e.g. parcels in a cadastre where the norm is one person owns one parcel, it is easy to add links to show ownership of many parcels by one person, multiple people owning one parcel etc. Such changes can be made on the fly by the user as required, and there is no need for a data supplier to redesign databases to accommodate such changes.

RDF and triples are a way of defining a network as the triple <subject, predicate, object> defines two nodes (subject, object) and the link (predicate). Spatial data currently held in relational databases can be converted to triple stores and managed with software such as Fuseki. Current relational databases can be made into virtual triple stores as well. Triple stores can be queried using SPARQL (SQL for triple stores).

Importantly, each element of a triple can be a URI (or IRI for different languages), allowing further distribution of data and definitions. For example, if a predicate is called "near", the IRI can point to a location where the concept is defined. It may be the Euclidean distance between two points (spatial) or the distance between people in a family tree.

Of importance to the semantic web, RDF enables access to knowledge and rules, as well as the data allowing sophisticated user defined operations to occur, again without the data supplier having to configure systems specifically for a user. Ontologies and rules allow high level queries and processing to occur by many users on the fly, which is currently not possible.

RDF was originally considered as metadata but now covers data as well. RDF triples can be used to represent tables, graphs, trees, ontologies and rules because it describes the relationship between subject and object resources where a 'object' in the <subject, predicate, object> triple can be another subject enabling subjects to be linked together. Each of the triple components can also be a URI so information can be linked across the Web. RDF formatted data is much easier to process, because its generic format contains information that is clearly understandable as a distributed model.

Reasoning and rules are an important part of this research and in the Semantic Web, the Ontology Web Language (OWL-2), based on RDF, is used for defining Web ontologies that include rules, axioms and constraints allowing inferencing (discovery of new knowledge) to be performed.

The Semantic Web has been used for queries by a user for natural events using observation sensor data [7,8]. In particular [7] describe a number of ontologies used to model various sensors and rules used to map queries such as flooding in an area to the need to sample a number of point water sensors. Methods have been proposed that have potential to automate land development approval processes. For example, the Sensing Geographic Occurrences Ontology (SEGO) model supports inferences of institutionalized events [9] based on time. However they do not resolve any conflicts arising if an event qualifies based on both policy and business rules. This research does not cover the sensor-specific technical details [9], but instead concentrates on the business knowledge rules.

A large number of open source and proprietary tools are available for semantic web research and development. This research uses the Protégé framework (http://protege.stanford.edu/) to develop ontologies and rules because its GUI environment allows fast design, interactive navigation of the relationships in OWL ontologies and visualization. It allows some rule-based analysis to be performed and can read and write RDF-based files in a number of different formats. Rules are defined in the form of ontological vocabularies using Semantic Web Rule Language (SWRL). Like many other rule languages, a SWRL rule has the form of a link between antecedent and consequent. The antecedent refers to the body of the rule, consisting or one or more conditions, and the consequent refers to its head, typically one condition. Whenever the conditions specified in the antecedent are satisfied, those specified in the consequent must also be satisfied [10]. Once ontologies and rules have been defined, they can be imported into the Apache Jena framework complete with the Pellet reasoner (http://clarkparsia. com/pellet/) to support OWL for runtime querying and analysis [11]. Combining both Jena and OWL API libraries, Pellet infers logical consequences from a set of asserted facts or axioms.

3   Case Study

Landgate administers all official naming actions for Western Australia under the authority of the Minister for Lands. The relevant local government authority generally submits all naming proposals for ratification by Landgate. All new proposals must satisfy government policies and standards. The current process has an online submission form, but for the most part the process is paper-based and requires significant human involvement. Current methods often require negotiation between the parties involved (i.e. local government and Landgate). While there are specific rules applying to new road name approvals, there are grey areas within policy that are often challenged and can only be resolved by an experienced negotiator. A request for a new road name may be transferred back-and-forth until an outcome is achieved that is satisfactory to both parties. Outcomes may be different depending on the expertise of the negotiator/approver.

Automation is needed to reduce the manual overhead by extracting expert knowledge for road name approvals to create a standard set of rules. The notion is to create a self-service online mechanism for developers to submit new road names for approval, underpinned by a complex rule-base and querying process. Complexity comes from the flow on effect of such changes. A new land development results in a change to the surrounding road network. This has a flow on impact to property street addressing and an administrative boundary change.

The case study uses the Landgate geographic road names database, called GEONOMA, to process the road name proposal. The current online submission process has the following issues that complicate the approval process:

- The online form is only used to test whether new road names are allowable based on a set of road names that have been reserved for use. If a proposed name is a reserved road name then the request will fail. There is no opportunity to contest the decision.
- A maximum of ten names per application is allowed; meaning separate applications are required for larger subdivisions. It is not possible to conduct cross-reference checks against other submissions and therefore the process is open to error.
- The current system does not consider the spatial extent of roads. Figure 1 shows a schematic submitted for road name approvals that does not represent the actual proposed location of roads. Roads do not actually meet up; they are stylized with solid and dashed lines with arrows etc. Manual editing and digitising is therefore necessary to extract the full topology of the proposed road network complete with coordinates of junctions.
- The current system does not permit checks on phonetics and this is an issue for similar sounding names (e.g., Bailey, Baylee, Bayley, Baylea). Similar or 'like' names (e.g. Whyte and White) are not allowable under policy guidelines as they can cause confusion for applications such as emergency services

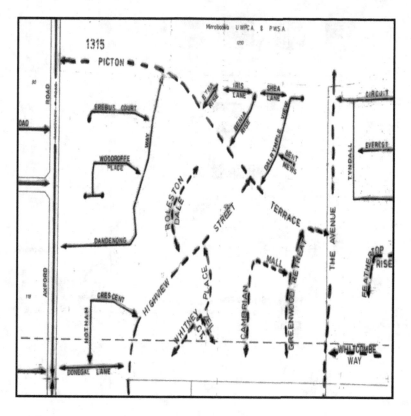

Fig. 1. Hardcopy road network plan with road name application.

dispatch. Similarly, the same road name or a similar sounding road name is not permitted within close proximity.

- Where an extension to an existing road occurs or where a road 'type' (e.g. cul-de-sac, highway) changes, the current system is unable to return an extension to a road name or change to road suffix, respectively.

4   Approach

Figure 2 shows the different phases in the land transaction process from knowledge acquisition to final feedback. Data is extracted from the various databases in formats such as html, JSON, csv and xml and converted to RDF. Ontologies in OWL are created from database schema and models in the interactive GUI based Protégé environment. Rules are generated in SWRL by an expert. Once the system has been developed, the data, ontologies and rules can be used in the runtime environment Jena with a rule engine by a developer to process road changes.

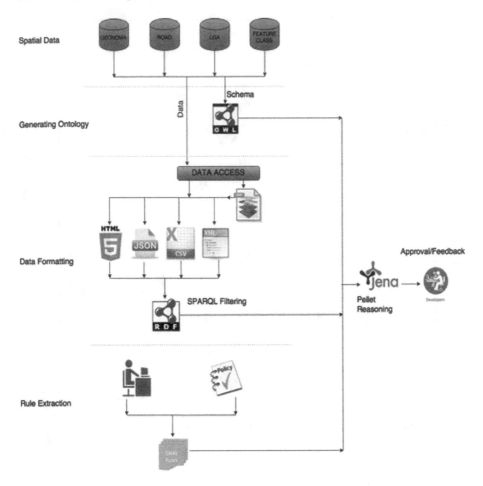

Fig. 2. Data integration/reasoning architecture.

4.1   Knowledge Acquisition

Knowledge acquisition was used to extract, structure and organise knowledge from policy documents, data dictionaries and by interviewing subject matter experts. This knowledge was then used to create the road naming rules. A combination of knowledge acquisition methods are used including organising explicit knowledge and eliciting tacit knowledge.

1. Organising explicit knowledge
   General procedures for spatial transactions are mentioned in policy documents, standards and dictionaries. These documents were reviewed to build the general rules on process. Establishing rules from explicit knowledge uses the following strategies:
   (a) Rules sourced from policy standards:

- A road name cannot be used if it already exists within a 10 Km radius of the new road in city areas or 50 Km in rural areas.
- A road name may not be used more than 15 times in the State of Western Australia.

(b) Rules sourced by accessing data dictionaries:

- Discriminatory or derogatory names are not allowed.
- A name in an original Australian Indigenous language will be considered for a new road name with reference to its origin.

2. Eliciting tacit knowledge

Currently polices and standards do not completely capture the human knowledge required for geographic naming processes. This makes it difficult to translate procedural knowledge into a computer-understandable form. In order to overcome this problem, knowledge elicitation techniques have been used to elicit procedural knowledge by conducting interviews, focus groups and observations etc.

(a) Rules sourced by interviewing subject matter experts:

- A name must not relate to a commercial business trading name or non-profit organisation
- A name must not sound like an existing name
- A name with the suffix type 'place' or 'close' cannot be assigned to a road greater than a specified length (200 m)
- A historical name, such as ANZAC, cannot be used
- A name with road type 'rise' can only be used for roads that have elevation or are at an incline
- Abbreviated names derived from the suburb name are not acceptable for new road names

With the current traditional naming process, satisfying the rules identified above is time consuming because of the back-and forth process between developer and approver. As an example, from a process perspective, when a land developer or local authority requests a new road name within a development site, a spatial validation process is run to test whether the proposed name:

- is already in use in the local authority and if so, whether it is within 10 Km of the new site; and
- has already been used 6 times within metropolitan area and 15 times across the State.

In addition to policy rules, subject matter experts use broader contextual knowledge when determining if a new road name is valid. For example, during the approval process experts check the scope for the proposed subdivision within the wider development site to avoid subsequent changes resulting from incorrect initial decisions.

Figure 3 presents a further example of where expert knowledge in the road naming process, from initial application to final approval, is required. During the negotiation phase with the land developer, documents are transferred back and forth between both parties; each making changes to a paper plan by way

Fig. 3. Road naming process in Jindalee - City of Wanneroo Western Australia. (Color figure online)

of communication. The following notes, written by Landgate to the developer, illustrate typical negotiations (See Fig. 3):

- Jindee Avenue: The road type is suitable, however the name Jindee is not. Apart from sounding similar to the suburb name, this is also an abbreviated name derived from the suburb name and is not acceptable. A replacement name is required.
- Limestone Street and Twinfin Way: The street is continuous so one street name can be used for this street.
- Noserider Drive: The name is suitable, however the road type Drive is not (as this road is adjacent below in this case) to a future open space then relevant types are Way, Vista, View, or if it shaped like a crescent, then Crescent can be used).
- Longboard Lane: The name complies with policy, however it is too long a word for that road. Also a portion of the extent is a part of Hilltop Lane (mentioned in green). A short name with its origin is required. Alternatively, the developer can hold the name Longboard for future use when a long road name is needed in the vicinity.

- Lifesaver Lane: the name is suitable, however it appears that there will be a third entry off Twinfin Court. Clarification of this will be necessary and an additional name for a portion (i.e. the northern east/west portion) will be needed.
- Midsummer Avenue and Treat Street: extensions are suitable because there are possibilities for the future development. The roads on the south side of Jindee Avenue (A & B) are currently unnamed as they are part of a later development stage.

4.2   Ontology Development

Ontology is one of the technologies listed within the Semantic Web Technology Stack [12]. Although it is used within the information sciences the term ontology has its origin in philosophy and is the study of being or existence [13] and it has been considered to be a branch of metaphysics looking at the nature of being. It is from these origins that the disciplines of Computer Science and Information Science borrow ontology and now it is used as a way to represent knowledge [13].

The term ontology is used with various different meanings and at different points in time these different definitions can be contradictory01 [14]. Bergmen [15] listed more than 40 different terms that are used which could all be called types of ontologies or at least ontological frameworks. With this number of terms often used in reference to ontologies it is quite understandable that there may be misunderstandings as well as misinformation about ontologies. Table 1 shows some of the various names that could loosely mean ontology. It is crucial that when using the term ontology it is clearly laid out how it is being used. Within

Table 1. Terms used to describe ontology (http://www.mkbergman.com/374/an-intrepid-guide-to-ontologies/).

Tag cloud	Social bookmarking	Topic Maps
Controlled vocabulary	Tags	Concept Maps
Thesauri	Tagging	Synsets
Collaborative tagging	Taxonomy	Glossary
Folk taxonomy	Folksonomy	WordNet
Directory	Classification	Data Reference Model
Subject Map	Categorization	Facets
Semantic Web	RDF	Structure
Cladistics	Metadata	Dublin Core
Markup languages	OPML	Typology
Ontology	XOXO	OWL
Microformats	Subject Trees	Information Architecture
Data dictionary	Phylogeny	

the work here within this paper the term ontology is used to describe the spatial aspects of land data and extract the rules to handle the decision making process.

Once the rules behind both policy standards and business processes are understood, the next step is to generate the ontology model from multiple sources of information. This ontology is developed as a global schema that means that while it works with the Landgate GEONOMA database, it can also be used in conjunction with other databases that link the spatial extent of a road to the road naming process. Figure 4 presents an overview of the generated Geo_feature ontology containing classes, data and object properties, and instances. Links show relationships such as domain, range and subClassOf. The ontological components are summarised below.

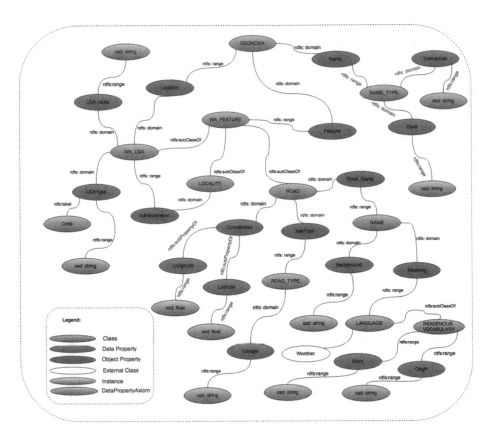

Fig. 4. An overview of Geo_feature ontology.

Geo_feature Ontology. The GEONOMA dataset is exported to XML and then imported into Protégé to help with the ontology generation process. Protégé was chosen, as it is an open source tool with wide community support that supports ontology development and reasoning, and importantly OWL DL, W3C description logic standard. The Geo_feature ontology consists of OWL classes, data and object properties, and individuals and is expressed in the form of OWL-2. Each OWL class is associated with a set of individuals. Object properties link individuals of one class to other class individuals. Data properties link one individual to its data values. Value constraints and cardinality constraints are used to restrict the attributes of the individual. For example each ROAD instance much have only one ROAD_TYPE through an object property link. Figure 5 shows the relationships between class instances. An example for a ROAD_TYPE instance is shown at bottom right. It has property restrictions handled by cardinality constraints. Each instance must have information about its type, description and whether it is a cul-de-sac or an open ended road type. Typically, further work is required to create the full semantics in the ontology. `Geo_feature ontology` comprises of more than one ontology such WordNet ontology and homophone ontology. All semantic relationships (links) between data components are needed because mapping from datasets directly is not adequate to explain the full model [16]. For example, every instance of ROAD, LGA and LOCALITY has a link with an instance of GEONOMA. Similarly every ROAD has a link with LGA and LOCALITY. These are inferred in Protégé by invoking the OWL-DL rule reasoner.

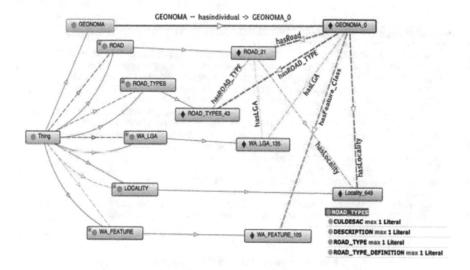

Fig. 5. OntoGraf representation for classes and instances.

Ontological Classifications and Spatial Relations. The resulting `Geo_feature ontology` represents the spatial relationship between several datasets including the road network, local government authority boundaries, locality and language. These datasets combined are used in the road name approval process and checked for constraints. The spatial relationship distinction is mainly based on source datasets. However, from a realistic viewpoint, these source datasets can only supply certain details relating to a feature name. To make it more meaningful there is a need to add additional vocabularies such as the Australian indigenous language dictionary and the WordNet ontology. The Australian indigenous language dictionary gives insight into the Australian indigenous naming specifics and WordNet ontology resembles a thesaurus of English words. By adding these we can check the meaning of a name and whether or not it complies with the chosen road-naming theme. To process a road request the road structure needs to be examined. By adding road coordinates it is possible to check where the proposed road will be actually developed.

4.3 Rule Development

Figures 4 and 5 show several relations between spatial datasets, such as the link between road and locality. Many of these relationships are inferred by the rule-based mechanism automatically from constraints, axioms and links defined in the ontology, thereby reducing the need for manual specification for all instances. The Pellet reasoner is used to infer decisions from these SWRL rules in Protégé. These inferred decisions are then communicated to the developer as a feedback. More complex, nested conditions can be handled by Boolean operators in SWRL rules are executed with the rule engine [17].

```
<owl:NamedIndividual rdf:about="http://www.semanticweb.org/ontologies/LandInfo#GEONOMA_0">
    <rdf:type rdf:resource="&GEONOMA;GEONOMA"/>
    <GEONOMA:FEATURE_NUMBER rdf:datatype="&xsd;integer">100062011</GEONOMA:FEATURE_NUMBER>
    <GEONOMA:NAME_TYPE rdf:datatype="&xsd;string">Approved Name</GEONOMA:NAME_TYPE>
    <GEONOMA:FEATURE_STATUS rdf:datatype="&xsd;string">CURRENT</GEONOMA:FEATURE_STATUS>
    <GEONOMA:DISPLAY_NAME rdf:datatype="&xsd;string">Chancery CT</GEONOMA:DISPLAY_NAME>
    <GEONOMA:FULL_NAME rdf:datatype="&xsd;string">Chancery CT</GEONOMA:FULL_NAME>
    <GEONOMA:SECURITY_CLASS rdf:datatype="&xsd;string">Complete Access</GEONOMA:SECURITY_CLASS>
    <GEONOMA:LOCALITY_NAME rdf:datatype="&xsd;string">Forrestfield</GEONOMA:LOCALITY_NAME>
    <GEONOMA:LGA_NAME rdf:datatype="&xsd;string">Kalamunda, Shire of</GEONOMA:LGA_NAME>
    <FEATURE_CLASS rdf:datatype="&xsd;string">Public Road</FEATURE_CLASS>
    <CATEGORY rdf:datatype="&xsd;string">Road</CATEGORY>
    <hasLGA rdf:resource="&LGA;WA_LGA_135"/>
    <hasFeature_Calss rdf:resource="&Ontology1445038673;WA_FEATURE_105"/>
    <hasRoad rdf:resource="&Ontology1445238488;ROAD_21"/>
    <hasLocality rdf:resource="http://www.semanticweb.org/ontologies/LandInfo#Locality_649"/>
    <hasROAD_TYPE rdf:resource="http://www.semanticweb.org/ontologies/LandInfo#ROAD_TYPES_43"/>
</owl:NamedIndividual>
```

Fig. 6. Source data in RDF format.

4.4 Data Formatting/Conversion

Once the ontology and rules have been developed the next stage is to access the source datasets to reason with the ontologies. To make this happen it is necessary to convert the source dataset into RDF triple format. In this way all data are accessible in one common format and ready for initial reasoning [18]. There are many data conversion and integration tools (Karma, MASTRO, OpenRefine and TripleGeo) that can be used for this conversion. MASTRO has been shown to be a successful Ontology-Based Data Access (OBDA) system through a series of demonstrations [19–23]. It can be accessed by means of a Protégé plugin. The facilities offered by Protégé can be used for ontology editing, and functionalities provided by the MASTRO plugin can be used to access external data sources. Openrefine (http:// openrefine.org/) is used to convert data to RDF format. Spatial information from a shape file can converted into RDF triples [24] (https://github.com/GeoKnow/ TripleGeo). Figure 6 shows an RDF instance. Having the data instances in RDF format, Apache Jena, with the help of MAVEN repositories is used to link all the ontologies, instances and rules at runtime.

5 Process/Operation

5.1 System Implementation

Figure 7 shows the runtime system architecture, which has been implemented using Jena in Java. The ontology repository consists of multiple ontologies

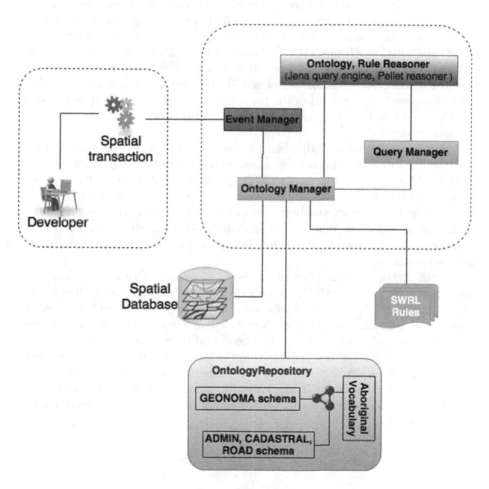

Fig. 7. System architecture.

derived from the data schema, data individuals, and rules, as well as non-specific ontologies such as Aboriginal vocabularies. The event manager collects the land transaction information and supports the ontology manager to infer the information relevant to that application. For example, if the application relates to a new subdivision, then it will gather the details spatially related to that land area, or if the proposed road name relates to a road name change, then it will gather information related to naming from the policy. The Ontology Manager collates the land information from the spatial database into the knowledge base.

5.2   Reasoning

The initial stage of reasoning is carried out in Jena with the Pellet OWL reasoner that checks the logical consistency of the model, processes the individuals (current, approved and proposed roads), infers new information including links and relationships, and updates the model with the inferred information. Through consistency checking, the system confirms whether or not any contradictory facts appear within the ontology. For example, the domain and range constraints on the feature relation: GEONOMA Features: `Feature_Class`. Constraints on the relation mean that GEONOMA has features, which come under only one of the `Feature_Class` categories. The reasoner will throw relevant errors if any ontological inconsistency appears given the proposed roads, for example if an instance of GEONOMA is linked to an instance of a ROAD and missing any property restriction relations.

Similarly, assigning an individual to two disjointed categories such as LGA and Locality will make the ontology inconsistent. Consider the case where every GEONOMA instance is represented with the ROAD feature type; it must have at least two coordinates and link to other road instances. This is declared as a necessary and mandatory condition for instances of the ROAD category in the OWL class description. When an individual in OWL satisfies such a condition then the reasoner automatically deduces that the individual is an instance of the specified category.

As well as the reasoning described above, to gather more information additional reasoning is required. Rules are expressed in terms of ontological vocabularies using SWRL. Table 2 shows some examples of implemented rules. As mentioned earlier, in each rule, the antecedent refers the body of the rule and the consequent refers to the head. The head and body consist of a conjunction of one or more atoms. Atoms are stated in the form of C(?R) P(?R,?X), where C and P represent an OWL description and property, respectively. Variables representing the individuals are in the form, for example ?R, where the variable R is prefixed with a question mark. Table 2 shows some examples of rules related to the application.

Table 2. SWRL rules with the action of each of the rules.

No	Purpose	SWRL rules
1	Relate a road link with existing road either directly or thru another proposed road	NEWROAD(?N), ROAD(?R), HASROADLINK(?N, ? R), STATUS(?N, "New"), STATUS(?R, "Existing"), notEqual(?N, ?R) − > isAllowed(?N, true)
		NEWROAD (?N1), NEWROAD (?N2), ROAD (?Old), HASROADLINK (?N1, ? N2), HASROADLINK (?N2, ?R), notEqual(?N1, ?N2), notEqual(?R, ?N2), STATUS (?R, "Existing"), STATUS (?N1, "New"), STATUS (?N2, "Aproved"), − > isAllowed(?R1, true)
2	Find the road duplication within the radios	NEWROAD(?N), ROAD(?R), ROAD_NAME(?N, ?RN2), ROAD_NAME(?R, ?RN1), stringEqualIgnoreCase(?RN2, ?RN1) − > hasRoad(?N, ?R)
3	Check the road name with definite article	DEFINITEARTICLES(?D), NEWROAD(?R1), FULL_NAME(?D, ?DN), ROAD_NAME(?R1, ?RN), stringConcat(?MSG, "Road name cannot contain definite article", ?DN), stringEqualIgnoreCase(?RN, ?DN) − > isAllowed(?R1, ?MSG)
		DEFINITEARTICLES(?D), NEWROAD(?R1), FULL_NAME(?D, ?DN), ROAD_NAME(?R1, ?RN), startsWith(?RN, ?SW), stringConcat(?MSG, "Road name cannot contain definite article", ?DN), stringConcat(?SW, ?DN, "") − > isAllowed(?R1, ?MSG)
4	Check for similar sounding names	NEWROAD(?N), ROAD(?R), LGAS(?N, ?L2), LGAS(?R, ?L1), METAPHONE_ALTERNATE(?N, ?MN2), METAPHONE_ALTERNATE(?R, ?MN1), ROAD_NAME(?N, ?RN1), ROAD_NAME(?R, ?RN2), containsIgnoreCase(?L1, ?L2), notEqual(?RN1, ?RN2), stringEqualIgnoreCase(?MN2, ?MN1) − > maySoundLike(?N, ?R)
		NEWROAD(?N), ROAD(?R), LGAS(?N, ?L2), LGAS(?R, ?L1), METAPHONE_ALTERNATE(?N, ?MN2), METAPHONE_PRIMARY(?R, ?MN1), ROAD_NAME(?N, ?RN1), ROAD_NAME(?R, ?RN2), containsIgnoreCase(?L1, ?L2), notEqual(?RN1, ?RN2), stringEqualIgnoreCase(?MN2, ?MN1) − > maySoundLike(?N, ?R)
		NEWROAD(?N), ROAD(?R), LGAS(?N, ?L2), LGAS(?R, ?L1), METAPHONE_ALTERNATE(?R, ?MN1), METAPHONE_PRIMARY(?N, ?MN2), ROAD_NAME(?N, ?RN1), ROAD_NAME(?R, ?RN2), containsIgnoreCase(?L1, ?L2), notEqual(?RN1, ?RN2), stringEqualIgnoreCase(?MN2, ?MN1) − > maySoundLike(?N, ?R)
		NEWROAD(?N), ROAD(?R), LGAS(?N, ?L2), LGAS(?R, ?L1), METAPHONE_PRIMARY(?N, ?MN2), METAPHONE_PRIMARY(?R, ?MN1), ROAD_NAME(?N, ?RN1), ROAD_NAME(?R, ?RN2), containsIgnoreCase(?L1, ?L2), notEqual(?RN1, ?RN2), stringEqualIgnoreCase(?MN2, ?MN1) − > soundsLike(?N, ?R)
5	Check the road name against road type	NEWROAD(?R1), ROAD_NAME(?R1, ?RN), ROAD_SUFFIX(?R1, ?RT), stringEqualIgnoreCase(?RN, ?RT) − > isAllowed(?R1, "Road name cannot be the same as road suffix")
6	Check the road length to against road types	NEWROAD (?R1), ROAD_SUFFIX(?R1, ?RT), hasLength(?R1, ?200), SameAs (?T1, ?*Close*) − > isAllowed(?R1, true)
7	Check the road name with restricted words	ILLEGALWORDS(?I), NEWROAD(?R1), FULL_NAME(?I, ?IN), ROAD_NAME(?R1, ?RN), startsWith(?RN, ?SW), stringConcat(?MSG, "Road name cannot contain word", ?IN), stringConcat(?SW, ?IN, "") − > isAllowed(?R1, ?MSG)
		ILLEGALWORDS(?I), NEWROAD(?R1), FULL_NAME(?I, ?IN), ROAD_NAME(?R1, ?RN), stringConcat(?MSG, "Road name cannot contain word", ?IN), stringEqualIgnoreCase(?RN, ?IN) − > isAllow

- Rule 1 automatically infers information with the help of a road link between proposed and existing roads from the source dataset with reference to road coordinates and feature id. This rule is necessary as every road needs to link with at least one other road to allow access.
- Rule 2 checks the similar road names within the neighbouring LGA to avoid duplication of road names.
- Rules 3 prevents the definite article being used in the road name.
- Rule 4 checks for similar sounding names within the LGA and neighbouring LGAs to avoid confusion for first responders and visitors to the locality.
- Rule 5 checks the road name against its road type to avoid road naming as road suffix.
- Rule 6 checks road length against road type. Checking the road length for shortest road types ('Place', 'Close' and 'Lane') is necessary to avoid confusion with the preference for road usage.
- Rule 7 prevents the restricted words such as 'CITY', 'SHIRE' and 'TOWN' being used in the road name.

Fig. 8. Automatic spatial transaction application portal.

Fig. 9. Automatic spatial transaction feedback.

6   User Interface

The automated spatial transaction application has been developed in this research using Jena in Java. Firstly, the user interface was designed to obtain input from the end-user and, secondly, the rules for geographic naming were built using SWRL and then linked with the Jena rule engine. The Jena engine is used to link all the ontologies, instances and rules at runtime with the help of MAVEN repositories.

The user interface allows the developer to select the development site from the map layout. From the selected site the system buffers either 10 Km or 50 km radius depending on the location of the site. Figure 8 shows the user interface for road naming transactions. Once the developer selects the development site the application then allows the developer to enter new road details. In many cases the development site will require the approval of several new roads. For this reason, the application provides an upload facility for developers to lodge a CSV file format to save time. The system is designed so that the road names contained in the CSV file are assessed simply by pressing the evaluate button. If any of the given road information does not comply with the rules, then the application provides feedback to the user accordingly. An example is shown in

Fig. 9. Once all submitted roads comply with the rules, then the system requests the developer's details and all supporting documents as evidence for further land development proceedings.

7   Conclusion

Traditional methods in spatial transaction mainly involve manual assessment of applications that cause delay, as a consequence of a back and forth process is being required. Human involvements are very time consuming, expensive and may trigger errors. This emphasized the importance of automation that reduces the manual overhead by extracting expert knowledge for such critical spatial transactions.

This paper proposes a Semantic Web solution for automating the decision making process for spatially related transactions. Examples of such transactions are approvals for new roads names and property address change. The method develops a `Geo_feature ontology`, which comprises knowledge of roads and constraints, axioms and rules extracted from sources such as experts, policy, geometry and past decision documents. The method shows how ontologies and rules are manipulated with reasoning techniques to infer new information.

Semantic Web techniques are used as the solution because it allows the ontologies and rules to be published in RDF and made available for other application domains. For example, similar processing is envisaged for points of interest (bridges, parks), and the reconciliation of addresses. These ontologies can be used in other jurisdictions for similar transactions or other application domains.

This method has proven successful for the process that involves simple spatial queries, such as a request for new road name approval and updating existing road features. The User interface facilitates the developer and government agencies in naming proper road names by providing feedback with map layout that helps the developer to understand road name non-compliance faults in visual form. More rules and relationships with existing ontology elements are being developed as further examinations are carried out into the datasets and business rules. Future work is also examining reasoning over other information that can be used to aid the approval process. For example an approver may use aerial photography to check for the presence of vegetation, as the removal of trees may need approval, and digital elevation maps used to determine if the proposed roads are viable.

Acknowledgement. The work has been supported by the Cooperative Research Centre for Spatial Information, whose activities are funded by the Australian Commonwealth's Cooperative Research Centres Programme. The authors extend their thanks to Landgate for providing the example datasets for the case study and subject matter experts for rule formulation.

References

1. Varadharajulu, P., Saqiq, M.A., Yu, F., McMeekin, D.A., West, G., Arnold, L., Moncrieff, S.: Spatial data supply chains. Int. Arch. Photogramm. Remote Sens. Spat. Inf. Sci. **40**(4), 41 (2015)

2. Lee, B.T., Fischetti, M.: Weaving the web: the original design and ultimate destiny of the world wide web by its inventor (1999)
3. Gupta, S., Knoblock, C.A.: A framework for integrating and reasoning about geospatial data. In: Extended Abstracts of the Sixth International Conference on Geographic Information Science (GIScience) (2010)
4. McMeekin, D.A., West, G.: Spatial data infrastructures and the semantic web of spatial things in Australia: research opportunities in SDI and the semantic web. In: 2012 5th International Conference on Human System Interactions (HSI), pp. 197–201. IEEE (2012)
5. Shadbolt, N., Berners-Lee, T., Hall, W.: The semantic web revisited. IEEE Intell. Syst. **21**(3), 96–101 (2006)
6. Millard, E.: The semantic web could enable even greater access to information. Promise of a better Internet. Teradata Mag. online (2010)
7. Devaraju, A., Kuhn, W., Renschler, C.S.: A formal model to infer geographic events from sensor observations. Int. J. Geogr. Inf. Sci. **29**(1), 1–27 (2015)
8. Yu, L., Liu, Y.: Using linked data in a heterogeneous sensor web: challenges, experiments and lessons learned. Int. J. Digit. Earth **8**(1), 17–37 (2015)
9. Reitsma, F.E.: A new geographic process data model. Ph.D. thesis (2005)
10. O'Connor, M., Knublauch, H., Tu, S., Grosof, B., Dean, M., Grosso, W., Musen, M.: Supporting rule system interoperability on the semantic web with SWRL. In: Gil, Y., Motta, E., Benjamins, V.R., Musen, M.A. (eds.) ISWC 2005. LNCS, vol. 3729, pp. 974–986. Springer, Heidelberg (2005). doi:10.1007/11574620_69
11. Segaran, T., Evans, C., Taylor, J., Toby, S., Colin, E., Jamie, T.: Programming the Semantic Web. O'Reilly Media Inc., Sebastopol (2009)
12. Hazaël-Massieux, D.: The semantic web and its applications. In: World Wide Web Consortium, SIMO - The Semantic Web and its applications, November 2003
13. Gruber, T.R.: Toward principles for the design of ontologies used for knowledge sharing? Int. J. Hum. Comput. Stud. **43**(5–6), 907–928 (1995)
14. Noy, N.F., McGuinness, D.L., et al.: Ontology development 101: a guide to creating your first ontology (2001)
15. Bergman, M.: An intrepid guide to ontologies. ai3::: Adaptive information (2007)
16. Ghawi, R., Cullot, N.: Building ontologies from multiple information sources. In: 15th Conference on Information and Software Technologies (IT2009), Kaunas (2009)
17. Powell, J.: A Librarian's Guide to Graphs, Data and the Semantic Web. Elsevier, Waltham (2015)
18. Broekstra, J., Kampman, A., Harmelen, F.: Sesame: a generic architecture for storing and querying RDF and RDF schema. In: Horrocks, I., Hendler, J. (eds.) ISWC 2002. LNCS, vol. 2342, pp. 54–68. Springer, Heidelberg (2002). doi:10.1007/3-540-48005-6_7
19. Calvanese, D., De Giacomo, G., Lembo, D., Lenzerini, M., Poggi, A., Rodriguez-Muro, M., Rosati, R., Ruzzi, M., Savo, D.F.: The mastro system for ontology-based data access. Semant. Web **2**(1), 43–53 (2011)
20. Poggi, A., Rodriguez, M., Ruzzi, M.: Ontology-based database access with DIG-Mastro and the OBDA Plugin for Protégé. In: Proceedings of the 4th International Workshop on OWL: Experiences and Directions (OWLED 2008 DC), vol. 496 (2008)
21. Savo, D.F., Lembo, D., Lenzerini, M., Poggi, A., Rodriguez-Muro, M., Romagnoli, V., Ruzzi, M., Stella, G.: Mastro at work: experiences on ontology-based data access. In: Proceedings of DL, vol. 573, pp. 20–31 (2010)

22. Rodriguez-Muro, M., Lubyte, L., Calvanese, D.: Realizing ontology based data access: a plug-in for Protégé. In: IEEE 24th International Conference on Data Engineering Workshop, 2008 (ICDEW 2008), pp. 286–289. IEEE (2008)
23. Zhang, Y., Chiang, Y.-Y., Szekely, P., Knoblock, C.A.: A semantic approach to retrieving, linking, and integrating heterogeneous geospatial data. In: Joint Proceedings of the Workshop on AI Problems and Approaches for Intelligent Environments and Workshop on Semantic Cities, pp. 31–37. ACM (2013)
24. Patroumpas, K., Alexakis, M., Giannopoulos, G., Athanasiou, S.: TripleGeo: an ETL tool for transforming geospatial data into RDF triples. In: EDBT/ICDT Workshops, pp. 275–278 (2014)

A Spatial Ontology for Architectural Heritage Information

Francesca Noardo[✉]

Politecnico Di Torino – DIATI, C.so Duca Degli Abruzzi, 24, 10129 Turin, Italy
francesca.noardo@polito.it

Abstract. The evolved potentialities of information technologies permit data disambiguation, interoperability and sharing through the web to reach an effective comprehensive knowledge. International standards are published as a reference for integrating data in a common framework and in an open perspective. Standard ontologies exist both in the cartographic and the cultural heritage field; however, they are distinct standards, and some limits (in the spatial or semantic management) make them incomplete for being used to manage architectural heritage knowledge. It is necessary to exploit both disciplines' contributions, integrating them in a model suitable for architectural heritage data management. In this paper, the ontological model for cartographic urban themes, OGC CityGML, is extended for managing architectural heritage multi-scale, multi-temporal, complex data. The conceptual framework is explained and some implementation aspects are considered, both for the definition of the extension and for the filling-in of such a structure with architectural heritage 3D data.

Keywords: Semantics · Standard · Ontologies · 3D model · CityGML · Interoperability · ADE · Architectural heritage

1 Introduction

The management of spatial and geographic knowledge is becoming increasingly discussed, since informatics technologies permit newly-advanced analyses and possibilities in information sharing. Contextually, some connected principles and concepts highlight new requirements for knowledge and new needs for data management. In particular, interoperability is a key issue, upon which the ideas of the Semantic Web, smart cities and international standards are built [1–3].

A unique framework is therefore necessary to make conceptualisations unambiguous. This can be solved through the use of ontologies [4, 5] to reduce the risk of misinterpretation and possible consequent damages or loss of information [6]. Moreover, the definition of an explicit and shared data model permits the production and sharing of open data, with all the connected advantages [7]. Such known and explicit structures also permit performance of advanced analysis and enhanced queries, and artificial intelligence mechanisms can be exploited for extracting new knowledge. In fact, in addition to being the objects of the usual queries, datasets can be considered for inferring new knowledge, through both using simple deductive mechanisms and

© Springer International Publishing AG 2017
C. Grueau et al. (Eds.): GISTAM 2016, CCIS 741, pp. 143–163, 2017.
DOI: 10.1007/978-3-319-62618-5_9

applying more complex procedures that also perform inductive or abductive reasoning. Structured datasets can also be the basis of effective data mining. The difference between this and a simple database query is that users may not know what information or patterns they are seeking in advance [8].

The world of spatial knowledge management and geographical intelligence is therefore developing tools for the realisation of an effective "geoweb" [9]. We can see this effort in the directives of some national and international institutions dealing with cartography or environmental management. For example, the INfrastructure for SPatial InfoRmation in Europe (INSPIRE) European Directive was developed by the European Parliament and the Council on the 14th of March, 2007 (Directive 2007/2/EC) [10]. Similarly, some consortiums of major stakeholders and actors of the sector are developing international industry standards, becoming the base for interoperability and open data. Within this framework, the Open Geospatial Consortium (OGC) [11] standards (including the model for urban data CityGML) were conceived.

The OGC CityGML [12] is an open data model, available in the form of an application schema (XSD) for Geographic Markup Language (GML) files. It is aimed at the representation, storage and exchange of three-dimensional (3D) urban objects. The original aim of CityGML, beginning in 2007, was to foster the reusability of 3D city models. Its semantic definition can be equally useful for managing the semantics of data with the tools offered by informatics and artificial intelligence.

More examples can be seen in the opposite direction; that is, the effort of the world of semantic thematic data to include geographic information. For this reason, GeoS-PARQL [13] was developed by OGC as an extension of the World Wide Web Consortium's (W3C) [14] SPARQL Protocol and RDF Query Language (SPARQL) [15], which is the language designed to query Resource Description Framework (RDF)-structured data. It is considered for the inclusion of spatial data in RDF-Ontology Web Language (OWL) information.

These languages are defined by the cited organisations, and represent crucial technology for the application of the theories of open-data and interoperability. Among these, some markup languages allow content to be written and provide information about which role the content plays using both human and machine-readable formats. In particular, eXtensible Markup Language (XML) [16] is used as a metalanguage for markup, because it provides a uniform framework and tools for the interchange of data and metadata among applications. For this reason, XML is the base of most of the languages created to structure open and application-independent data, and exchange them through applications or through the web. Some relevant XML-based languages include RDF [17], which permits the management of semantic data (through a triple mechanism); OGC GML [18] for archiving geographical objects; and Collaborative Design Activity (COLLADA) [19], which is an interchange format for 3D models. The structure of XML-based files is defined in equally XML-based formats, such as simple XML Schema Definition (XSD), which is used by GML format, or extended ones such as RDF Schema (RDFS)-OWL [20].

An example of geographic issues managed on the web using the described technologies is GeoNames [21], which is a database including toponyms, gazetteers and information related to the included named places.

Looking at the cultural heritage (CH) field, database interoperability and information retrieval have always been crucial aims for documentation [22]. It is indispensable for data to be unambiguous, permitting a correct interpretation, and to be contextualised with metainformation.

The International Committee for Documentation (CIDOC) conceptual reference model (CRM), developed by the Committee of the International Council of Monuments, (ICOM) is considered the core ontology for cultural heritage [23]. It uses RDF-OWL for the management of thematic data and became the standard ISO 21127.

A further existing database exploiting the described theories and technologies is a set of vocabularies developed by the Getty Institute [24]. This set is oriented to structure cultural heritage related terms and items, and is divided into four vocabularies: the Art and Architecture Thesaurus (AAT), the Thesaurus of Geographic Names (TGN), the Union List of Artist Names (ULAN) and the Cultural Objects Name Authority (CONA). The AAT hierarchically structures terms linked to the description of works of art and architectures, whereas the Getty TGN, in contrast to GeoNames, also includes historical denominations. The ULAN contains names and synthetic information about cultural heritage authors; and finally, the CONA describes the different denominations of a cultural item over time. Within these vocabularies, the spatial component is not present, but they can be the reference for the denomination of parts which unequivocally have a spatial connotation (e.g. all architectural parts or toponyms) or for related information (such as authors or object names).

Recently, some effort has been made to also include geographic information in cultural heritage descriptions, with attempts being made to include some localisation data in semantic structures. For example, the Getty project Arches [25, 26], based on CIDOC-CRM structure, integrates some WebGIS function, and the CRMgeo project [27] includes spatio-temporal representation potentiality in CIDOC-CRM.

These geographic references are often bi-dimensional, however, and have little defined geometry (points, lines or approximate polygons), since the aim is not the analysis and reading of the artefact geometry, but also its localisation for a territorial reading. Recently, another extension of the CIDOC-CRM was created: CRMba, which was expressly realised for the documentation of standing buildings [28]. A gap in this research could be found in the management of complex 3D models in connection with other parts of the city and the landscape, however, which is a topic addressed by CityGML.

For the particular needs of architectural heritage information management, two-dimensional (2D; often small-scale) data are not sufficient. 3D dense data must be exploited with higher levels of detail (high measurement and georeferencing accuracies) and complex semantic definition (object-oriented structures) [29].

The availability of 3D dense models is a realised aim of survey and geomatics discipline [30]; however, the potentiality of management, analysis and editing, typical of traditional Geographical Information Systems (GIS), is at the present moment reduced for this kind of data. The development of new software structure or user interfaces is necessary, based on either adapted or new theoretical frameworks [31, 32], to once again permit the usability of the systems in a truly effective way.

1.1 Proposal Aims

In this study, a solution for the needed data model for architectural heritage 3D high-level-of-detail data is proposed, by extending the existing structure OGC CityGML using its application domain extension (ADE) procedure.

CityGML was chosen as a base since it is a standardised model already meant to deal with buildings in their double dimensions: as a part of the city and the landscape, and as a higher-detailed 3D object. It is important to also consider this double nature in architectural heritage emergences, because they are often both meaningful to the definition of the cultural values of the considered buildings. Moreover, CityGML includes the possibility to have multi-scale representations. The integration of the monument into wider maps of the city or the landscape permits the performance of strategic analysis in a broader context.

CityGML is shared as a data model, already in a potentially implementation-ready format. Unified Modelling Language (UML) diagrams are published in the OGC encoding standard [33], already in an advanced phase of the data modelling process because the database design details are specified as in a logic-level model (e.g. an object-oriented approach is envisaged and types of data and code-lists are defined). Moreover, XSD files are shared and available for direct use in implementation. Because of its generality in representing urban models, however, CityGML can be considered an ontology [34, 35]. It is, in fact, independent from the specific applications for which it can be used, and aims to represent a common frame for urban 3D maps data.

For extending CityGML, including structures for the management of spatial data complexity of architecture and monuments, some preliminary general reflections are therefore reported, which can be valid as ontological-level thinking. Nevertheless, the extension is then realised in accordance with the formats and structures used in CityGML (implementation-oriented), in order to be coherent with the extended model and to also permit testing in the implementation. Some considerations and necessities of representation remain at present unimplemented, however, specifically concerning more evolved constraints to be imposed upon the model.

In a second part, the procedure followed for the implementation of the model is presented, highlighting some possibilities for the use of the schemes for data archiving.

In the end, some parts dealing with the kind of data to be managed are presented, taking into consideration the processing phases to be followed (from the processing of the 3D model to its semantic visualisation).

2 CityGML Cultural Heritage Application Domain Extension (CHADE)

CityGML model can be extended to model further aspects linked to specific application domains. The so-composed extensions use specific characteristics and procedures of CityGML, being defined as Application Domain Extensions (ADEs). Some official ADEs exist [36], especially regarding some urban-scale themes, such as noise or inclusive routing. Some of these are specifically for buildings; for example, GeoBIM integrates some classes derived from Industry Foundation Classes (IFC) standard [37] used in Building Information Modelling (BIM) [38]. Even if in the future the field of BIM (born to project new buildings) will likely meet GML models, at present it is too rigid for describing cultural heritage buildings, which require more flexibility.

Further research has been performed for the extension of the CityGML model in order to include some information about the cultural heritage nature of the building and some surface characteristics, such as deterioration [39]. In the model proposed in this paper, including the characteristics of surface complexity is attempted. Moreover, some attention is drawn to the traceability of the stored information, in order to include useful elements in the data to allow technicians to interpret the stored information and evaluate the degree of fuzziness in the data.

The CityGML Cultural Heritage Application Domain Extension (CHADE) for the building module of CityGML is summarised in Fig. 1 and then analysed in detail in the following subsection. The extension has been developed and will be tested on the building module. Once its validity is proved, its concepts and classes can also be applied to the other CityGML modules.

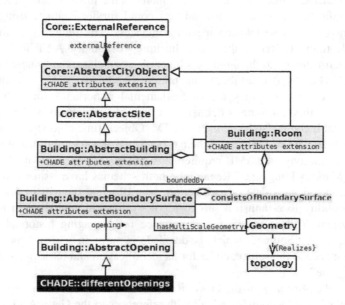

Fig. 1. Synthesis of the CityGML CHADE in a UML diagram. The CityGML classes are shown in white, the CHADE extensions and the inserted relations are shown in grey (black for the whole class).

2.1 CHADE Components: Research of Granularity, Flexibility and Traceability

From the general to the particular, the first problem was to include some attributes useful for the identification of the monument and some related information, such as if a CH declaration exists, what the related documents are, who the owners are and what the preservation authority is. Some of these have been borrowed from previous research [40] and extend the Core class "AbstractCityObject". It is possible to include this kind of extension by means of composite attributes in following phases; that is, the implementation of the model, since the XSD format permits the inclusion of complex attributes in the form of DataType, is composed of a series of further attributes. Another interesting possibility for this case is the "ExternalReference" class, already in CityGML, which permits the relation of the model with further databases managing data concerning the same object. For example, considering the management of the Versailles castle, the reference can be realised to the instance having ID:700000350 of the CONA vocabulary of the Getty Institute, which describes it (http://www.getty.edu/cona/CONAFullSubject.aspx?subid=700000350).

The second issue is the extension of the attribute list for the "AbstractBuilding" class; in particular, its function and its denomination. Both of these values are complex when regarding a historical item, since both can change over time, and must be archived both as a reference for research and as an element for understanding the history of the building. In the implementation phase, a DataType is therefore included for both. The BLDG_Function data type first includes the function name (at present in English, specific future works with historians could further evaluate using different languages in order to avoid losing meaning nuances). The reference to the Unique Resource Identifiers (URI) of the Getty Institute vocabulary AAT follows, which includes the terms linked to the building's function as subclasses of "single built works by function". The last two attributes are present nearly everywhere in the detailed added data types, because they are of fundamental importance for historical data connotation. The time attribute is defined as a time object defined in the same GML general schema. It could also be defined as a TM_Object (time object), as stated in ISO TC211 ISO 19108:2006 Temporal Schema, but some incompatibilities among some ISO TC211 definitions and GML requirements persist (https://en.wikipedia.org/wiki/Geography_Markup_Language). Regardless, both schemas have issues with detailing the time considered, as a date or as a period, with different degrees of fuzziness and with the possibility to establish a sort of topology for temporal data, in a temporal reference system. This is of obvious importance for managing historical data. The second attribute is "Source", which is detailed, in turn, in a data type, including metadata, reference to the source, codes for its identification and retrieval, and the same attribute of "time".

Similarly, the attributes of the class "Room" are extended, adding "RoomClass", "RoomFunction" and "RoomUsage", all with references to the Getty AAT vocabulary URI. "RoomUsage", which can change over time, is detailed in a dedicated data type.

Perhaps the more interesting part of the model is the extension of the CityGML class "AbstractBoundarySurface". In the original model, this had no attributes, and could be specialised as belonging to the main parts of the buildings (e.g. RoofSurface, CeilingSurface, WallSurface). The change in this class can permit very flexible descriptions of the parts of the buildings, which are stratified and articulated, and even small portions can have different meanings, with "portions" and "fiat parts" (without "bona fide", that is defined, boundaries) being intended. A recursive mereological "part of" relation is therefore added from AbstractBoundarySurface to the same AbstractBoundarySurface, which permits the articulation of the surfaces in hierarchical, semantically clear, multiscale and possibly topologically-related parts. Several attributes are added and defined following the previously explained criteria. Among these, the "LevelOfSpecialisation" (LOS) attribute deserves an explication. 3D models are usually considered for geometric accuracy, which mainly derives from the production methods and measurements systems. This characteristic is stored in GML models as Level of Detail (LOD) associated to each geometry. Even if it implies some consequence on the level of semantic definition, it is mainly linked to the possibilities of representation offered by the available data, and thus to the accuracy and data density. The LOS is added in order to manage the possibility to define parts and subparts that can be recognised on the same model (with homogeneous accuracy and LOD) but need to be separately specified because of the different meanings they assume if considered as part of the whole or as a singular part (Fig. 2).

Fig. 2. Examples of consecutive LOS specified on parts of a homogeneous-LOD 3D model. The colours represent the parts into which the object is divided.

A further extension of this class regards the associated geometric levels of detail; because the aim of CityGML is a cartographic representation, the envisaged levels of detail are not high enough to be used for small architectonic details. The CityGML LODs are conceived to manage data from an urban or regional scale (approximately 1:25,000) to a higher level of detail, useful for representing some building characteristics. Because CityGML is oriented to an urban representation, however, the maximum detail that can be included in the model, while remaining meaningful to the representation and consistent with the other levels, is approximatively 1:500 [41]. More detailed representation can be achieved using textures on the surfaces, permitting the addition of some details usually present in higher representation scales. To include architectural heritage in the urban representation, however, it is necessary to reach further levels of detail, because smaller parts should also be geometrically represented without losing their complexity. Two more LODs are therefore added: a LOD5, for approximately 1:200 or 1:100 scale, and a LOD6 for larger scales. The associated geometry class must be defined as a "Geometric complex::GM_CompositeSurface", since it is structured and hierarchical, in the same way as the boundary surfaces must be also semantically defined. Moreover, it must be related to a "Topological Complex::

TP_Complex", deriving from the "Topology" part of the standard ISO TC211 – ISO 19107:2003 Spatial Schema or the GML specification (they should be harmonised for the same issues regarding the time objects). This last described part is complex to use with current software and requires more effort. Regardless, the inclusion of the topological relations as schematised in Fig. 3 should be useful for correctly setting the models.

OGC has processed some topological and mereotopological structure to help correctly store data, but they are already oriented as linked open data formats, without regarding GML [42].

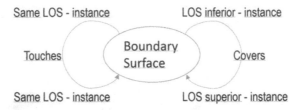

Fig. 3. Schema of Egenhofer Topological Relations, to be included in the model and verified. Some exceptions can exist (for example, a pillar can be considered only for one half as a component of a bay, admitted that they belong to the same classification), and these must be analysed in order to confirm or refute the validity of this model. Egenhofer relations are considered even if they deal with 2D geometry, because only a reference surface is considered, although it is a 3D surface.

2.2   Implementation Issues

Ontologies exploit object-oriented structures and methods, which are unusual in current and known GIS management systems; some of the most widely-used ones (PostgreSQL-PostGIS, ArcGIS Geodatabase tool, Oracle) implement object-relational schemas, which are hybrid systems including some constructs of object-oriented databases, but not the whole potentiality. Several studies concerning the development of semantic GIS have been performed, beginning in the mid-1990s [43, 44]. In these studies, an object-oriented approach was used as an effective solution for expressing and storing data meanings [45]. In this way, even more powerful systems could be built with significant data interoperability and a reduction of any potential ambiguity. Regardless, few similar systems are currently available, and SQL-based implementations are preferentially used [46]. This is due to the necessity to adapt the exigencies to the available platforms and software systems, and to change the storing methods to permit the production and management of computationally heavy files. In the near future, object-oriented GIS will likely be developed again, or a different interface from the GIS we know will be improved upon to include spatial analysis and query functionalities.

The described model has been implemented using the method defined as best practice by OGC [47]. UML schemas are modified, using stereotypes defined by a GML UML profile, so that their meaning can be understood by the machine and the performed transformation can be coherent and correct. In particular, for building the system, the proprietary commercial software Sparx Systems – Enterprise Architect is used as, contrary to the indications of using open source software for managing public (and open) data, it also is recommended for some official occasions (for example for the management of INSPIRE schemas, or in OGC practices). The software permits the import of existing models; in this case, obviously, the CityGML building module and some general schemas such as GML are used, although ISO 19108 for temporal objects and ISO 19107 for spatial issues could be considered. The classes, selected and imported in the new extension model, maintain all their characteristics and relations with the other parts of the model to which they belong. This is crucial to avoid creating an isolated new model, but rather to insert into a complex existing framework.

From this basis, new classes can be added, attributes can be defined and new relations can be established. In particular, following OGC best practice for extending an existing class with further attributes, a subclass having the same name as the class to be extended and stereotyped "ADEElement" should be created. The specialisation relation is marked with stereotype "ADE", whereas to add a new class, a simple subclass having stereotype "featureType" must be added [48]. The so-formed model (Fig. 4) can then be exported in different formats, including XSD, to be used as a GML application schema.

Other interesting formats included OWL and ArcGIS workspace. From the XSD file, Simple Query Language (SQL) relational or object-relational database schemas could also be generated. The transformation can be directly performed by the software, exporting the model in SQL format, to be used for generating an SQL database. The translation is not so immediate, however, because the codes and the data types used in the different languages vary, making it necessary to carefully control them. A further possibility is to generate an SQL database from the XSD by passing through different software, such as Altova XMLSpy, which permits the management of XML documents, and eventually manually adapt the data formats. The passage from an object-oriented model to a relational one often cause some complexity to be lost, however, and the analysis aims for which the ontological structures and data models were conceived are therefore undermined. For this reason, it is recommended to use GML data and XSD structures, although SQL data is more easily manageable, thanks to the well-affirmed and evolved software tools with which they can be managed.

Following the generation of the XSD for structuring complex GML data, the translation can be automatically performed from the modified UML model. The passages must be controlled and corrected, since they need correct information or reference files describing, in specific software-understandable language, how the transformation must be done. Some specific applications and a proper UML profile exist for this aim, but they are not always easily available and compatible in any situation. While waiting for this progress, possibly planned as future work, the resulting files of the transformation must be manually corrected by editing the XML text following the rules for the ADEs realisation.

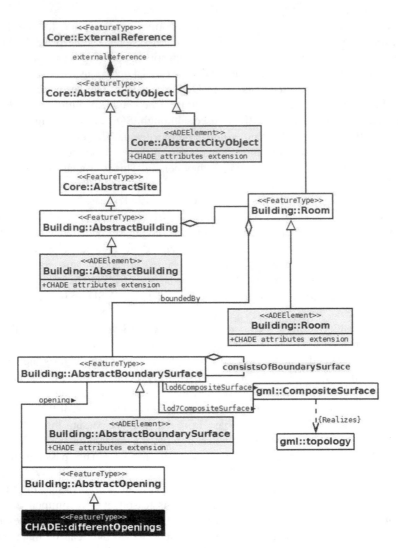

Fig. 4. Synthesis of the UML model modified for extending CityGML – Building module in the CHADE, following OGC best practice indications (Van den Brink et al. 2012).

3 Preparing the 3D Data

For processing and interchanging the data concerning the current case study, which is the medieval Staffarda Abbey (in the north-west of Italy), some of the previously-described technologies are used. The main problems throughout the whole workflow are often linked to the inability of the software tools to manage some functionalities, algorithms and/or formats simultaneously. For this reason, several passages using specific software are needed.

In this study, existing and available software are used, as the implementation of new applications was not among the objectives. In particular, open source solutions are preferred, when possible, for interoperability and replicability issues. Moreover, open source software can often be modified and adapted to the specific application, without being constrained to the existing structures.

When the schemas are ready, the dense high-level-of-detail 3D models must be prepared. Because the managed surfaces are complex, being composed by miles (or millions) of triangles (stored in the form of rings composing a multiple composite surface), it is obviously not possible to manually store the singular point coordinates, but they must pass through a series of phases which permit their export in a GML format.

The church was surveyed with a high level of detail using integrated geomatics techniques. The lower part of the church and the interior were acquired by means of a terrestrial laser scanning technique. The obtained model was integrated with the roofs and the higher part of the church, measured and modelled using images from an

Fig. 5. Views of the 3D model (textured mesh) of the Staffarda Abbey, processed using LIDAR acquisitions integrated with photogrammetric data acquired from an unmanned aerial vehicle [49].

unmanned aerial vehicle, and then processed in digital photogrammetry and structure-from-motion software [49]. The final model was highly accurate, because it is meant to be useful as support for a preservation project. Here, the acquisition and processing phases are not described in detail. In brief, these phases exploit a series of techniques for georeferencing the model in a known reference system [50, 51], for measuring points with various methods and variable accuracy and density [52], and for processing the models to finally obtain an integrated, correct, georeferenced and optimised 3D model (Fig. 5). The obtained model was further integrated to include another, even higher, level of detail, which could permit the analysis of the single parts of a capital. This model was obtained by means of terrestrial images, processed with digital photogrammetry and integrated with the structure-from-motion technique.

It is not necessary here to describe the process of the model's generation; we will instead start the process from the already generated, georeferenced and optimised 3D model of the church.

The first editing phases to be performed on the models regard two main processes, the first of which is the reduction of the model's points, while conserving the original definition (Fig. 6). This process can be performed using different algorithms, which are not always available in proprietary software. Regardless, the topic should be further analysed in order to establish the methods and the limits of this practice.

The second process involves segmentation of the model, isolating every part from the other parts to allow suitable management, both geometrically and semantically (Fig. 7). Moreover, temporal connotation must be considered in the segmentation, since it is essential for historical objects [53]. A surface can be eventually repeated if it is part of more than one instance having different LOS as attributes. Furthermore, a quantity of techniques should be analysed to find the most suitable one. In this first application, each part of the model was manually segmented in a 3D modelling software; Hexagon 3D Reshaper was used, but the operations could be equally realised using open source software, such as Meshlab or Cloud Compare.

Fig. 6. 3D models of the Staffarda Abbey façade before (I) and after (II) the reduction of the number of triangles composing the surface. Both model views are in wireframe modality (only the triangle edges are shown).

Fig. 7. 3D model showing texture of one segmented capital of the Staffarda Abbey.

These phases can be performed in 3D model processing and editing software, such as Hexagon 3D Reshaper (proprietary software), which offers the advantage of managing coordinates that have high values, such as cartographic ones, allowing georeferenced models to be directly managed. Moreover, the software possesses integrated advanced editing tools.

At this point in the process, two main alternatives are available for translating the 3D model into a CityGML-compliant format. The first is the use of Safe Software FME (again, a proprietary software), which is expressly dedicated to these operations. FME permits the mapping of objects in different formats and schemas to several predisposed data models, implemented in the software libraries. In particular, it is conceived for filling in CityGML databases or INSPIRE-compliant datasets, although it is possible to translate the original data to other specific formats. Because FME is a close proprietary software, however, it is impossible to modify its internal libraries to include the extended schema CityGML-CHADE. For this reason, manual editing of the resulting file is necessary to adapt the data to the new schema.

The second option is the use of ESRI ArcGIS, which is also a proprietary software; however, because it is widely-used, its formats and procedures are often considered to be de-facto standards. ESRI ArcGIS is used in this case for this reason, even though the passage from an ESRI shapefile format, which is based on a relational model, does not grant the opportunity to directly specify the final structure of the data. Additionally, the ESRI processing integrates FME algorithms through the ArcGIS "Data Interoperability Toolbox" extension.

For using the processing integrated in ESRI ArcGIS, more passages are necessary. The processed 3D model must be exported in COLLADA format for the following transformation. This open exchange format is not managed by some proprietary software; therefore, the model must be exported in a 3D model format (such as OBJ or PLY) and reimported into further software tools, such as the open source software MeshLab, able to conduct the exportation. The problem is that Meshlab has difficulties in managing high coordinate values, so that the whole model must be translated near the origin for this passage. The exported COLLADA files can then be reimported in ESRI ArcGIS, as multipatch shapefiles [54]. Here, they must be retranslated to their original position in georeferenced coordinates and can then be exported through the "Data Interoperability" toolbox to a generic CityGML file. The result is the inclusion of both the geometry and the attributes of the single parts of models in files structured as CityGML and semantically classified as "GenericCityObjects". The GML file (readable as XML structured text) must be manually modified to be included in the CHADE description schema and to correctly define the semantics of each part.

In particular, a reference to the extension namespace must be added in the heading of the file, since there is no way of modifying the FME libraries (used directly or through ArcGIS toolbox) to include the extensions. Moreover, each segmented part of the multipatch is exported as a distinct object having a geometry attribute (in the form of gml::MultiSurface), but the parts are not hierarchically structured and they do not yet have specific semantics (being all "GenericCityObjects"). The hierarchy must therefore be set and the correct labels must be applied following the CityGML file format. Moreover, all the textual attributes must be manually filled in. In this phase, it is applied the extended model CityGML+CHADE. Another important issue is the addition of suitable identifiers in order to uniquely identify the objects for performing queries, retrieving information, and realising certain connections. T the Xlink syntax (which requires IDs for linking to specific objects) is often used for this. It is preferable for the IDs to be composed in the form of URI, following the rules used in best practices for a linked data environment [47]. In this way, the produced information could be more easily translated to linked data for effective sharing and processing through the Semantic Web. The Xlink syntax can also be considered and used for the establishment of mereotopological relations among the parts.

XML processing software, such as the proprietary software ALTOVA XMLSpy or the open source software Xpad, can validate the obtained GML file.

4   Results: The Archive in a Graphical Interface

At this point, the GML file could be shared through the web and read by several applications or interfaces to allow consultation and analysis. In the current case, an open source software was used and tested in order to read the GML archive based on CityGML CHADE. An open source software was chosen for two main reasons: the previously-cited goals of interoperability and replicability of the procedures, and because open source software tools often permit access to the source code of the libraries they use, allowing possible modification. This is useful for including the CHADE schema for the correct interpretation of objects by which it is referred.

The FZK software (http://www.iai.fzk.de/www-extern/index.php?id=2315) was used, which is one of the more well-developed CityGML viewers available. It includes the schemas of some versions of CityGML, as well as some official CityGML ADEs (e.g. the Noise ADE). Furthermore, it has an open structure, which can be customised by adding, for example, other CityGML schemas to be used. For this research, the CHADE schema (in XSD) was added to the directory of the reference files of the software to allow the described data to be interpreted. This is unequivocally an advantage of the open structure of the software. In this way, the software is able to read the resulting GML archive (Fig. 8).

In the visualisation platform, the objects inserted in the archive can be read, including the relationships among them (Fig. 9). Some measurements can be directly made on the 3D model (Fig. 10) and some thematic visualisation can be generated, similarly to in GIS management software environments (Fig. 11). Moreover, statistics concerning the data are computed.

The application should be developed in order to include the possibility of effectively managing some elements introduced by the extension, however, such as the inclusion of different addresses (referring to the building, but also to the owners, the authority, etc.), which occasionally causes problems in their visualisation. In the same way, the links inserted in the GML file for cross-referencing the objects or for inserting references to external resources (for example the Getty vocabularies) do not function in the software, likely because some changes in the reading of such components should be made.

Additionally, the levels of detail that can be visualised are limited to the those envisaged by CityGML. To include more detailed levels added in the CHADE, the application should be modified not only by adding the schemas, but also by modifying its tools and interface code. Furthermore, the possible thematic visualisations are limited to some attributes of CityGML and do not consider those introduced by the extension. The same is true for the statistics and analyses that can be performed, which are limited to some pre-set parameters. It would be interesting to attempt to enhance the parameters. These limits are connected to the visualisation platform, however, whereas the previously structured GML file is independent from the platform.

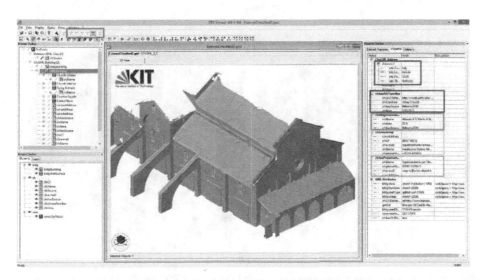

Fig. 8. GML model structured using the CityGML CHADE in the FZK software interface; on the left, the objects in the model are listed, in the center the 3D model is visualised and on the right, the properties can be read. The attributes, which are, in turn, objects themselves or data types (and are therefore composed of a set of attributes) are highlighted by the frames. The level of detail to be visualised can be chosen, since the data are multi-scale (in the left part of the toolbar, framed in the figure).

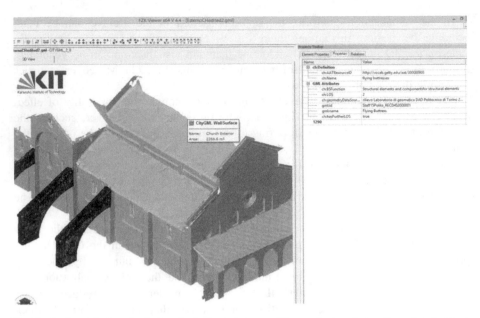

Fig. 9. In the right box ("relations" window), it is possible to select and visualise related objects (geometry and thematic attributes). The image on the left is the result of the relation of the whole object to one of its parts (the flying buttress). The specific parts are selected in the representation and the attributes are listed on the right side of the figure.

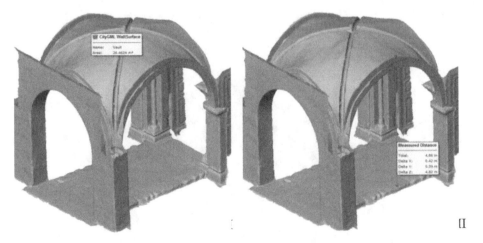

Fig. 10. Example of direct measurement possibilities on the 3D model: areas (I) and distances (II). This can be extremely useful for architectural heritage researchers and operators.

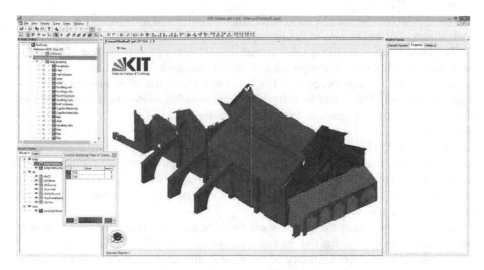

Fig. 11. Example of a thematic visualisation of the Staffarda Abbey (based on the attribute "year of construction").

5 Discussion and Perspectives

Reference domain ontologies are essential for managing information in the framework of interoperability established by recent theories developed by informatics and communication disciplines, which are shared with the Semantic Web and the world of standards. Architectural heritage information should also be effectively documented using this technology. There is no suitable model for architectural heritage documentation, however. For this reason, the available standards dealing with building, landscape or city representation, and cultural heritage management were analysed and taken as a starting point, either for developing an architectural heritage extension or for their reciprocal integration. The OGC CityGML model was selected as the best reference ontology for representing and documenting buildings. The CityGML standard data model can be considered an ontology independent from the specific application, although many implementation details are proposed and defined in the published standard.

Multi-faceted, complex, multi-scale 3D models require a specific structure to be effectively managed. This goal was realised by means of an extension of the OGC CityGML model, which was proposed and tested for this aim. The important aspects and issues typical of architectural heritage are included in the extension, from both the spatial and the thematic points of view.

Two main implementation phases should be considered: defining the data model extension and filling in the model with data. The first phase, that is, the conceptual definition of the extension, was implemented by means of some existing tools, which are proposed in the same standards as best practice for realising extensions. Some residual difficulties in correctly and automatically performing some passages remain, however; the resulting products must therefore be manually refined by editing the generated XML file to obtain a valid XSD. Major control and automatisation in process performance will be the aim of future improvements.

The same limits in realising automatic processes are found in managing 3D models, for which complex steps are required and passage through different software tools to prepare the data is often necessary. The generated GML file must also be manually edited in this case. This could be solved if it were possible to modify the software tool libraries by inserting extensions to the reference models, as little alternative exists to their use.

The use of XML language as a base for writing the files is a definite advantage, because it can be read by humans and by a number of applications, permitting reading, processing and editing. Moreover, using XML also requires basic tools (even a simple text-editor could be effective) to be managed. Conversely, the required skills are not within everyone's reach.

In the development of the model, some fundamental aspects are addressed, such as the granularity of the information and its traceability, which is essential when dealing with historical items; the flexibility of the model, allowing adaption to the representation of such unique artefacts as monuments; and the inclusion of thematic data with eventual reference to external databases and vocabularies.

Some automated reasoning based on the assembled knowledge is enabled by the realisation of standardised datasets using reference ontologies; this is even more powerful if the information is shared and accessible on the web. Furthermore, the use of ontologies permits the interoperability of databases and easier retrieval of information in compliant databases, including the Semantic Web. Architectural heritage research and monument preservation issues can obviously gain great advantages by using such systems, but these systems can also be critical for connected topics, such as administration, tourism, risk analysis and other activities.

At present, the constructed archive can be used in applications similar to the known GIS, which permits surfing of the archive, realisation of queries, application of symbols, measurement of the model, and so on. The available platforms should be modified and improved, however, in order to permit a wider range of analyses and statistics, and to include enhanced visualisation options.

Future work will be performed to include true management of advanced spatial issues, such as topology and mereotopological constraints, in the models and in the data, in order to enhance analysis potentialities and transversal information retrieval.

A further improvement will be necessary to make the connection from external reference to vocabularies effective (possibly using methods similar to the use of gazetteers for toponyms). Additionally, an inclusion or link to further data models and ontologies will be considered; for example, a connection to the CIDOC-CRM could be of primary importance.

Another critical passage will be the translation of both the model and the dataset as linked open data. This will permit effective exploitation of Semantic Web technologies and connection to similar information. This translation will be a future development of this study and proposal.

Once these aspects are solved, a further step towards world-wide management of architectural heritage documentation in a unique effective framework for their preservation, retrieval and analysis will have been made.

References

1. Barnaghi, P., Wang, W., Henson, C., Taylor, K.: Semantics for the Internet of Things: early progress and back to the future. IJSWIS **8**(1), 1–21 (2012)
2. Chourabi, H., Nam, T., Walker, S., Gil-Garcia, J.R., Mellouli, S., Nahon, K., Scholl, H.J.: Understanding smart cities: an integrative framework. In: 2012 45th Hawaii International Conference on System Science (HICSS), pp. 2289–2297. IEEE (2012)
3. Schaffers, H., Komninos, N., Pallot, M., Trousse, B., Nilsson, M., Oliveira, A.: Smart Cities and the future internet: towards cooperation frameworks for open innovation. Future Internet Assembly **6656**, 431–446 (2011)
4. Guarino, N., Oberle, D., Staab, S.: What is an ontology? In: Staab, S., Studer, R. (eds.) Handbook on Ontologies. IHIS, pp. 1–17. Springer, Heidelberg (2009). doi:10.1007/978-3-540-92673-3_0
5. Laurini, R.: Geographic ontologies, gazetteers and multilingualism. Future Internet **7**(1), 1–23 (2015)
6. Guizzardi, G.: Ontological foundations for structural conceptual models. CTIT, Centre for Telematics and Information Technology (2005)
7. Janssen, M., Charalabidis, Y., Zuiderwijk, A.: Benefits, adoption barriers and myths of open data and open government. Inf. Syst. Manage. **29**(4), 258–268 (2012)
8. Worboys, M.F., Duckham, M.: GIS: A Computing Perspective. CRC Press, Boca Raton (2004)
9. Laurini, R.: A conceptual framework for geographic knowledge engineering. J. Vis. Lang. Comput. **25**(1), 2–19 (2014)
10. INSPIRE European Directive. http://inspire.ec.europa.eu/. Accessed 11 Nov 2015
11. Open Geospatial Consortium. www.opengeospatial.org/. Accessed 11 Nov 2015
12. OGC CityGML. http://www.opengeospatial.org/standards/citygml. Accessed 10 Nov 2015
13. OGC geoSPARQL. http://www.opengeospatial.org/standards/geosparql. Accessed 12 Nov 2015
14. World Wide Web Consortium. www.w3.org. Accessed 09 Nov 2015
15. SPARQL. http://www.w3.org/TR/rdf-sparql-query/. Accessed 12 Nov 2015
16. XML. www.w3.org/XML/. Accessed 09 Nov 2015
17. RDF. www.w3.org/RDF/. Accessed 10 Nov 2015
18. GML. www.opengeospatial.org/standards/gml. Accessed 12 Nov 2015
19. COLLADA. https://it.wikipedia.org/wiki/COLLADA. Accessed 12 Nov 2015
20. OWL. www.w3.org/2004/OWL/. Accessed 11 Nov 2015
21. Geonames. http://www.geonames.org/. Accessed 12 Nov 2015
22. Restoration charters. http://www.icomos.org/en/charters-and-texts. Accessed 11 Nov 2015
23. Doerr, M., Ore, C.E., Stead, S.: The CIDOC conceptual reference model - a new standard for knowledge sharing. In: Tutorials, Posters, Panels and Industrial Contributions at the 26th International Conference on Conceptual Modeling, ACS, vol. 83, pp. 51–56 (2007)
24. Getty vocabularies. http://vocab.getty.edu/. Accessed 12 Nov 2015
25. Getty ARCHES project. http://archesproject.org. Accessed 09 Nov 2015
26. Myers, D., Avramides, Y., Dalgity, A.: Changing the Heritage inventory paradigm, the ARCHES open source system. Conserv. Perspect. GCI Newslett. **28**(2), 4 (2013)
27. Doerr, M., Hiebel, G., Eide, Ø.: CRMgeo: linking the CIDOC CRM to geoSPARQL through a spatiotemporal refinement. In: Institute of Computer Science, Technical report GR70013 (2013)
28. Ronzino, P., Niccolucci, F., Felicetti, A., Doerr, M.: CRMba a CRM extension for the documentation of standing buildings. Int. J. Digital Libr. **17**(1), 71–78 (2015)

29. Laurini, R., Thompson, D.: Fundamentals of spatial information systems. Academic Press, London (1992)
30. Chiabrando, F., Spanò, A.: Points clouds generation using TLS and dense-matching techniques. A test on approachable accuracies of different tools. ISPRS Ann. Photogrammetry, Remote Sens. Spat. Inf. Sci. **5**, 67–72 (2013)
31. Brahim, L., Okba, K., Laurini, R.: Mathematical framework for topological relationships between ribbons and regions. J. Vis. Lang. Comput. **26**, 66–81 (2015)
32. Solovyov, S.A.: Categorical foundations of variety-based topology and topological systems. Fuzzy Sets Syst. **192**, 176–200 (2012)
33. OGC. CityGML UML diagrams as contained in CityGML Encoding Standard Version 2.0, OGC Doc. No. 12-019 (2012)
34. Métral, C., Falquet, G., Cutting-Decelle, A.F.: Towards semantically enriched 3D city models: an ontology-based approach. In: Academic Track of GeoWeb (2009)
35. Kolbe, T.H., Gröger, G., Plümer, L.: CityGML–3D city models and their potential for emergency response. In: Zlatanova, S., Li, J. (eds.) Geospatial Information Technology for Emergency Response. CRC Press. (2008)
36. CityGML ADEs. http://www.citygmlwiki.org/index.php/CityGML-ADEs
37. IFC. https://en.wikipedia.org/wiki/Industry_Foundation_Classes. Accessed 09 Nov 2015
38. de Laat, R., van Berlo, L.: Integration of BIM and GIS: the development of the CityGML GeoBIM extension. In: Kolbe, T.H., König, G., Nagel, C. (eds.) Advances in 3D Geo-Information Sciences. LNG&C, pp. 211–225. Springer, Heidelberg (2012)
39. Costamagna, E., Spanò, A.: CityGML for architectural heritage. In: Rahman, A.A., Boguslawski, P., Gold, C., Said, M.N. (eds.) Progress in Cultural Heritage Preservation. LNG&C, pp. 219–237. Springer, Heidelberg (2012)
40. Costamagna, E., Spanò, A.: Semantic models for architectural heritage documentation. In: Ioannides, M., Fritsch, D., Leissner, J., Davies, R., Remondino, F., Caffo, R. (eds.) EuroMed 2012. LNCS, vol. 7616, pp. 241–250. Springer, Heidelberg (2012). doi:10.1007/978-3-642-34234-9_24
41. Fan, H.C., Meng, L.Q.: Automatic derivation of different levels of detail for 3D buildings modelled by CityGML. In: 24th International Cartography Conference, Santiago, Chile, pp. 15–21 (2009)
42. Geographical linked open data. http://ows10.usersmarts.com/ows10/ontologies/. Accessed 11 Nov 2015
43. Mennis, J.L.: Derivation and implementation of a semantic GIS data model informed by principles of cognition. Comput. Environ. Urban Syst. **27**, 455–479 (2003)
44. Fonseca, F., Egenhofer, M., Davis, C., Camara, G.: Semantic granularity ontology-driven geographic information systems. AMAI Ann. Math. Artif. Intell. **36**(1–2), 121–151 (2002)
45. Scholl, M., Voisard, A.: Object-oriented database systems for geographic applications: an experiment with O2. In: Proc. Int. Workshop on Database Management Systems for Geographical Applications. Springer. (1992)
46. Belussi, A., Liguori, F., Marca, J., Migliorini, S., Negri, M., Pelagatti, G., Visentini, P.: Validation of geographical datasets against spatial constraints at conceptual level. In: UDMS 2011: 28th Urban Data Management Symposium, Delft, The Netherlands, 28–30 September 2011. Urban Data Management Society, OTB Research Institute for the Built Environment, Delft University of Technology (2011)
47. Van den Brink, L., Janssen, P., Quak, W., Stoter, J.E.: Linking spatial data: automated conversion of geo-information models and GML data to RDF. Int. J. Spat. Data Infrastruct. Res. **9**, 59–85 (2014)

48. Van den Brink, L., Stoter, J.E., Zlatanova, S.: Modelling an application domain extension of CityGML in UML. In: ISPRS Conference 7th International Conference on 3D Geoinformation, The International Archives on the Photogrammetry, Remote Sensing and Spatial Information Sciences, 16–17 May 2012, Québec, Canada, vol. XXXVIII-4, part C26. ISPRS (2012)
49. Bastonero, P., Donadio, E., Chiabrando, F., Spanò, A.: Fusion of 3D models derived from TLS and image-based techniques for CH enhanced documentation. ISPRS Ann. Photogrammetry, Remote Sens. Spat. Inf. Sci. 2, 73–80 (2014)
50. Chiabrando, F., Lingua, A., Piras M.: Direct photogrammetry using UAV: tests and first results. In: ISPRS International Archives of the Photogrammetry, Remote Sensing and Spatial Information Sciences, vol. XL-1/W2, pp. 81–86 (2013). ISSN: 2194-9034
51. Dabove, P., Manzino, A.M., Taglioretti, C.: GNSS network products for post-processing positioning: limitations and peculiarities. Appl. Geomatics 6(1), 27–36 (2014)
52. Bryan, P., Blake, B.: Metric Survey Specifications for English Heritage. English Heritage, Swindon (2000)
53. Donadio, E., Spanò, A.: Data collection and management for stratigraphic analysis of upstanding structures. In: Proceedings of the 1st International Conference on Geographical Information Systems Theory, Applications and Management, pp. 34–39 (2015). doi:10.5220/0005470200340039, ISBN 978-989-758-099-4
54. ESRI: the multipatch geometry type. ESRI White Paper (2008). https://www.esri.com/library/whitepapers/pdfs/multipatch-geometry-type.pdf. Accessed 12 Nov 2015

The Software Design of an Intelligent Water Pump

Daniel Scott Weaver$^{(\boxtimes)}$ and Brian Nejmeh

Department of Computer and Information Science,
Messiah College, Mechanicsburg, PA 17055, USA
sweaver@messiah.edu

Abstract. In an effort to increase handpump reliability, the Messiah College Collaboratory and the Department of Computer and Information Science are developing the Intelligent Water Project (IWP) system. IWP measures and reports the functionality of handpumps and volume of water extracted on two-hour intervals daily. Additionally, IWP will measure groundwater levels which can be used to evaluate well yields. Data from handpumps is automatically collected and transmitted to a remote database. Once in the database, the data is analyzed and distributed to stakeholders via web and mobile applications and customizable alerts. Besides monitoring water extraction, handpump performance, and borehole health, the IWP system processes data to alert stakeholders of failure or degrading conditions (imminent failure). Coupled with appropriate field management processes, this information can lead to improved handpump availability and lowered cost of ownership. The key goal is to dramatically increase the reliability of handpumps. A secondary goal is the collection of handpump data from all IWP enabled pump sources providing a rich resource of data to enabling WASH practitioners, managers, hydrologist and donors to make more informed decisions. The purpose of this paper is to highlight the key IWP software design considerations and to discuss the key software design decisions made and the rationale for the same.

1   Introduction

Wells and handpumps in Africa fail at alarming rates within the first two years of installation [1]. Much of this failure can be attributed to a lack of transparency into the performance of handpumps. Existing manual methods of handpump monitoring require manual field inspection by personnel which is costly, untimely and superficial. Furthermore, traveling long distances to reach handpumps results in infrequent inspections.

The advent of low cost, reliable sensor technology coupled with the ubiquitous GSM network has the potential to bring unprecedented levels of transparency to handpump performance in rural Africa. Our project has refined a fully automated wireless, sensor-based mobile and web application suite to provide significant remote transparency of the overall handpump performance.

© Springer International Publishing AG 2017
C. Grueau et al. (Eds.): GISTAM 2016, CCIS 741, pp. 164–180, 2017.
DOI: 10.1007/978-3-319-62618-5_10

Initial concept development of the Intelligent Water Project (IWP) sensor technology and the software suite began in 2012 with internal funding from the Messiah College Collaboratory and the Department of Computer and Information Science (CIS) with subsequent funding from World Vision. This project is being done by a coordinated group of faculty members and students across various engineering and computer science disciplines at Messiah College, a Christian college based in the United States. The software for this project has been developed using the Agile Scrum method [2] in two service-learning computer science classes (database applications, senior capstone course in CIS). Given the Christian-faith tradition of Messiah College, we often use Biblical references to help motivate our work. Work on the IWP project has been inspired by the following passage: *"The poor and needy seek water, but there is none, their tongues fail for thirst. I, the Lord, will hear them; I, the God of Israel, will not forsake them. I will open rivers in desolate heights, and fountains in the midst of the valleys; I will make the wilderness a pool of water, and the dry land springs of water."* Isaiah 41:17–18 (NKJ).

A system-level overview and results to date of the IWP, including the overall hardware and communications design, were chronicled in a prior paper [3]. This prior paper also contained a broad review of related work. The focus of this paper is on the software-specific design considerations and decisions made in the IWP.

IWP is differentiated from other handpump monitoring systems because it:

1. provides support for automated, sensor-based handpump data collection over the ubiquitous GSM network,
2. provides full transparency and access to all of the underlying sensor data via the website,
3. supports configurable, periodic status alerts on user defined events of interest,
4. leverages the work of the Messiah College India MKII and Afridev Sustainability Studies that gives unique insight and focus to the sensor design [4],
5. provides full integration with Google Maps® and ESRI (GIS cloud environment) systems,
6. is a cloud-based application suite which runs on desktops and mobile devices,
7. is being developed by an interdisciplinary team of hydrologists, mechanical engineers, electrical engineers and computer scientists.

2 Key Requirements

2.1 Problem Statement

Approximately 184 million people living in Africa depend on handpumps for their daily water supply [5] with an estimated 50,000 new handpumps shipped to Africa each year [6]. Despite efforts to improve rural water service delivery, handpumps serving rural communities often fall into disrepair. According to data compiled by Rural Water Supply Network (RWSN) [1] from 20 African nations covering 345,071 wells in 2009, 36% of handpumps are non-operational. This results in a loss of capital investment in infrastructure and a negative impact on rural communities. When a community handpump breaks down, families are forced to find alternative water sources. Alternative sources may include carrying water a greater distance from a handpump in a neighboring community, or less protected sources such as hand dug scoops or surface water. The latter sources carry increased risk of water born disease. The increased time and energy spent collecting water and the potential for illness detract from more economically empowering activities.

Logistical challenges and costs hamper effective and efficient handpump monitoring and evaluation efforts in rural areas. To determine the condition of a handpump, water authority representatives must travel to each handpump location and perform a manual inspection. This process can result in lengthy down-times and high labor and transportation costs incurred by the community and/or sponsoring NGO or government organization. As a consequence, handpumps may go weeks without necessary repairs and Water and Sanitation Hygiene (WASH) managers are forced to make critical program decisions on incomplete data.

Given the critical importance of clean water, it follows that an accurate, reliable and low-cost tool to assess handpump performance efficiency and effectiveness would be valuable to many stakeholders. Improved handpump transparency can lead to better visibility and early warning of handpump problems. This will enable timely handpump remediation, thereby leading to improvements in overall pump efficiency and effectiveness in service to rural African communities.

2.2   Solution Overview

The primary goal of IWP is to develop a system to automatically capture and organize data about handpump functionality and performance from both sensor and human sources. This allows the IWP to alert stakeholders via web and mobile app, email, and text messaging of pump failure or degrading conditions. Coupled with appropriate field response processes, the information the system provides can lead to improved handpump availability with a lower cost of ownership. A secondary goal is the collection of handpump data from all IWP enabled pump sources providing a rich resource of data to enabling WASH practitioners, managers, hydrologists and donors to make more informed decisions.

The IWP team decided on the following design goals and desired outcomes to drive our process:

Design Goals

1. Design a solar-powered, GSM-enabled, pump monitor with an array of sensors to communicate with a cloud-based database application,
2. Design a web-based application suite to produce actionable information about handpumps,
3. Design a mobile app that exploits location-aware and other mobile capabilities for local field technology workers.

Desired Outcomes

1. Improved visibility of handpump performance and ease of maintenance and reporting,
2. Improved understanding of water extraction for each handpump (how much and when),
3. Improved understanding of well water level fluctuations,
4. Single, unified source for storage, access, and analyses of handpump related data.

3   Architecture

The IWP remotely monitors handpumps, including the Afridev and India MKII, through the use of an embedded monitor installed in the handpump. The monitor, connected to and collecting data from concurrently installed sensors, is equipped with a GSM modem to communicate with the cloud-based database application via text messaging through an SMS receiver service. The cloud-based database application parses the transmitted data, populates the database and determines the performance and status of the handpump (See Fig. 1).

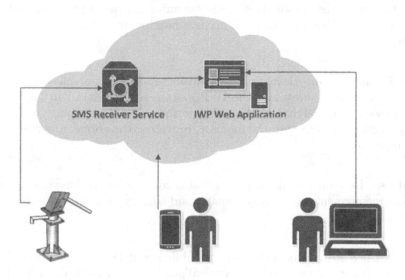

Fig. 1. The conceptual overview of IWP.

Each day the IWP embedded monitor measures and records the volume of water extracted by the handpump in two hour intervals, the amount of effort required to prime the pump, and the largest recorded leakage rate. This automatically collected data is transmitted daily to the remote database where the information is analyzed and made available to stakeholders via a web or mobile app. In addition to monitoring water extraction, handpump performance, and borehole health, the IWP system processes this data to predict certain degrading conditions before failure occurs and notifies stakeholders via customizable alerts. For instance, an increasing amount of effort required to prime the handpump may indicate degradation of handpump parts while decreasing yield may indicate silting at the screened interval.

In the event of an immediate handpump failure or degrading condition, the system automatically generates email and/or text message notifications to community members and area handpump technicians, mobilizing them to inspect the handpump and make repairs. Certain known failure modes are detected and reported by IWP, enabling repair teams to carry the needed parts, supplies, and tools to the site. Community members and area handpump technicians will also have the ability to report data such as cost of repairs or other visible handpump problems using the mobile app. Once a handpump is repaired, sensor data will verify handpump performance and close the failure reporting loop.

IWP is segmented into five architectural areas each with design goals and outcomes, providing team focus.

Monitor. Development of robust hardware powered by the sun that collects sensor data in 2-hour increments, formats a GSM text message with valid JSON formatting from the sensor data, and transmits the message to the database application. The Monitor, mounted inside a handpump, must be uniquely identifiable.

Data Transmission. Develop a mechanism to allow communication between uniquely identified pumps and the IWP system so that data can be collected to be viewed and analyzed in the web interface. The design must include the ability to communicate with the monitor, providing the ability to retransmit or dequeue messages.

Database. Develop a database model that is configurable and flexible to handle different requirements of different climates and landscapes from which the pumps send data.

Web Application. Inform the general public about IWP through blog entries and allow authorized users to access critical pump data through a mobile friendly web interface. Develop an intuitive graphic user interface through which users select pumps or pump groups via a geographic map and select data and graphs representing the sensor data from the pumps.

Mobile Application (Mobile App). Provide field technicians the ability to connect to the IWP system while in the field with recognizing that technicians will be in areas where 3G connectivity is unavailable.

4   Design Considerations

The main software subsystems of IWP include the monitor hardware, database, data transport, mobile app, and web interface. This section of the paper highlights the key design considerations for each of the IWP software subsystems.

4.1   Monitor Hardware

The IWP system hardware consists of a solar-powered, GSM cellular-enabled sensor node that mounts inside of India MKII and Afridev handpumps. A prior paper [3] details the hardware and mechanical design of IWP on the India MKII platform. The system monitors the motion of the handpump handle and the presence or absence of water in the mouth of the rising main to (a) ascertain the amount of upstroke required to prime the pump, (b) the amount of water extracted from the pump and (c) the rate of leakage in the rising mains. This information is summarized and sent daily to the IWP database.

The IWP monitor is embedded software that runs on a processor mounted inside of the handpump. It manages the collection, summary calculations and formatting of the sensor data. In essence, the monitor software collects, processes and formats the sensor data, thereby packaging the sensor data for transmission to the IWP application. There were several key software design decisions made with respect to the monitor:

1. global pump identification,
2. data granularity,
3. data format,
4. data structure.

The global name space for pump identity proved to be a significant design consideration. Given the hierarchical nature of pump identification among organizations using pumps (i.e., country, region, local community identifier), at first we thought we could leverage such information and create a hierarchical name space. However, such names were often quite lengthy in characters and could contain non-ASCII characters. A second option was to simply use the phone number of the SIM card used by the monitor to transmit the sensor data. However, a complication of using such a phone number is that it is entirely possible to relocate a SIM card to another pump for data transmission. Doing so would mean that the same SIM card could be associated with more than one pump over time. We also felt it was better for the global pump identity to be a property of the pump itself and nothing else. For this reason, we used a system generated global unique numerical identifier for global pump identity.

A second major design decision concerned the granularity of sensor data to be managed by IWP. For example, should we record the volume of water extracted at a pump every hour, every eight hours or every twenty-four hours? One could ask a similar question about the leakage of water within a pump. In the end, we decided to store the following data about a pump on a daily basis:

1. the daily average of water lost through leakage when the rising main is full of water,
2. the daily longest amount of upstroke required before water starts to be extracted (i.e., so-called "longest prime"),
3. estimated volume of water extracted (adjusted for leakage) per 2 h interval,
4. average extraction upstroke per pumping event.

A third significant software design consideration involved the format used to store and transmit the sensor data. We wanted a format that was an industry-standard and commonly used. The format also needed to be efficient (i.e., little data header and other overhead). In the end, we decided to use JSON [7], a widely used industry standard attribute-value format for our sensor data. In short, JSON met all of our data format criteria.

Finally, consideration had to be given to the data structure used to store and transmit the sensor data. Given the attribute-value pair nature of JSON, we decided to use a simple one character attribute identifier for each sensor data attribute followed by the value of the data for that attribute. Given that near-time sensor data analysis and reporting was deemed sufficient by field water specialists we consulted, we decided that we could transmit all of the data associated with a sensor over a 24 h period (i.e., one day) in a single JSON message. In addition to the date, one additional data item was included in the JSON message. In order to determine if each JSON message transmitted was successfully received and processed by the IWP application, we added a unique message sequence number to each JSON message. The next section of the paper will discuss the data transport design and how the unique message sequence number was used to guarantee delivery of all JSON messages to the IWP application.

In the future, we will consider adding additional sensors, and in turn, sensor data to the IWP. This will be easily accomplished by extending the JSON attribute-value format structure to incorporate the additional sensor data. Finally, consideration will be given to using a binary encoding of the JSON objects to provide an extra level of security and greater capacity to transmit data in a single JSON object.

4.2   Data Transport

The data transport software layer is responsible for the successful transmission of the JSON encoded sensor data over SMS. The data transport layer software design is depicted in Fig. 2. The Sensor data collected by the monitor is stored on both an SD Card installed in the monitor and resident memory. Every twenty-four hours the monitor packages the collected data as a JSON formatted string and creates an SMS message. The data transmission occurs at a slightly different absolute time (after midnight local time) on all monitors to reduce any possibility of a transmission bottleneck from occurring. The system transmits the SMS message over a GSM (voice grade) wireless network to an SMS Receiver Service. It is commonly known and field studies in Africa have shown [8] that voice grade

GSM network service is much more widely available than 3G network service. Given the desire to field IWP in remote areas of Africa, the decision was made to only assume a voice-grade GSM network in our design.

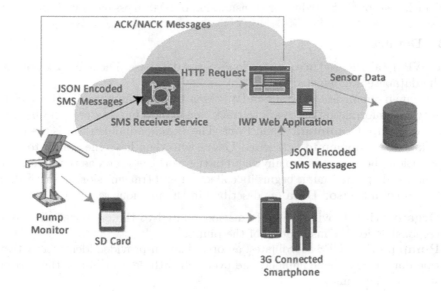

Fig. 2. Data transport.

The SMS Receiver Service forwards the message to the parser module in the cloud-resident database application. The parser module posts the raw message in the database, parses the message and (if error-free) stores the parsed sensor data in the database.

These SMS messages are equipped with unique sequence identifiers (USI). Once a message has been successfully processed and added to the database, an ACK message is sent to the monitor. Once received, the monitor will remove the message from its queue based on the USI.

If a message is missing data or the data does not conform to the defined data format, an ERR message is sent to the monitor. The monitor will then attempt to re-transmit the message.

When a duplicate message is encountered, a DUP message is sent to the monitor. In response, the monitor deletes the message from the queue.

There are several important future software design considerations for the data transport layer of IWP. To date, we have partnered with Upside Wireless® [9] to obtain a reliable SMS Receiver Service. In order to reduce costs and have greater flexibility, we are exploring the possibility of standing up and self-managing an open source SMS Receiver Service. Secondly, in some locations 3G+ wireless network service is reliably available. Such a wireless service would allow for the data transport layer to be IP-based and not require the use of SMS, but instead have the JSON data structure directly connect to the listening data parser on the server. Furthermore, such IP-based monitor software would allow for secure and direct monitor-based access to the sensor data on a more near-time basis (if that were desirable). Finally, consideration should be given to the digital replenishment of SMS credits for transmission of SMS messages by the monitor.

4.3 Database

The IWP data is housed in a secure, cloud-based database. The main components of the database are depicted in Fig. 3.

When an organization installs an IWP Monitor in a handpump within a community, the information necessary to track that handpump is stored in **Organization**, **Pump**, **Community**, and **Part**. The system administrator is then able to link that handpump to authorized **Users** who then have the ability to view information about that pump anywhere in the world. As soon as the Monitor is installed and operational, it begins its collection and transmission of sensor data to be stored in **Sensor Data** (as described in the previous section).

1. **Organization** provides the information related to the organization that is responsible for the installation of the pump.
2. **Pump** provides GPS coordinates, an organization-provided identifier, activation date, GSM phone number, and overall health data including the current status of the pump.

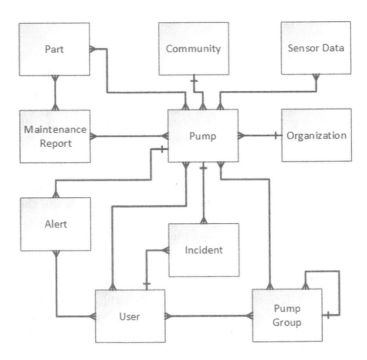

Fig. 3. Database conceptual model.

3. **Community** provides the information related to the community in which the pump resides including the community name and country.
4. **Part** provides information about the parts contained in the pump along with field maintenance records for replacement parts including commission and decommission dates, cost of part, and other maintenance-related information.
5. **User** provides the information for authorized users of the system including personal information, access level, and pumps they are authorized to monitor.
6. **Sensor Data** is stored both as un-parsed JSON strings as well as parsed JSON strings, providing access to all sensor collected data.

The insertion of sensor data into the database triggers a process that calculates health indicators, such as leakage rate and, based on configurable parameters, determines the current status of the handpump. If the status of the pump changes, the system creates an **Incident** report (SIR), which in turn generates an **Alert**. The user, having the capability to indicate how he/she would like to be notified, will receive the alert on login to the web or mobile app, email, or via text message.

Any authorized user may also create an **Incident** report based on their observation of the handpump. These human-generated incident reports (HIRs) are included in the calculation of the pump status. **Alert**s are generated based on the worse of the two types of incident reports (HIRs or SIRs).

1. **Incident** is divided into two categories: System Incidents and Human-recorded Incidents. System Incidents contain information related to the calculated health indicators of the pump. There are methods to customize what thresholds are used to change the status of pumps. Human-recorded incidents are those field technicians may enter via the mobile app. These are combined with System incidents to determine the overall health of the pump.
2. **Alert** is used to provide messages of health changes in a pump. The system is designed to allow authorized users to indicate the frequency of the sending of alert messages as well as the desired device through which the message is received.

IWP provides a mechanism for grouping handpumps together using **Pump Groups**. Pump groups allow for aggregate analysis, reporting, and searching across pumps in that group. Pumps can belong to more than one group. A pump group has a unique name and a brief description and allows pumps to be grouped by location, field technician, type of pump, or any user-defined grouping.

Pump groups were designed to provide an organizational tool for users. In the future, the implementation of global pump groups will allow users to easily view their assigned pumps within those groups. For example, pumps within the borders of a country would be assigned to that country pump group and users would be able to see their assigned pumps within that country without creating their own grouping.

An additional future consideration will be to automatically generate pump groups based on attribute-pair configuration parameters. For example, pump groups may be generated with all handpumps within an x-kilometer radius of a given GPS location, or those assigned to a given field technician, or within a certain geographic boundary such as a country.

When handpump technicians perform maintenance on a handpump, they complete a **Maintenance Report**, identifying their incurred travel, part, and labor costs. As mentioned above, maintenance reports also include **Pump** and **Part** data, providing a mechanism for calculating the cost of ownership for any given pump.

Security Model. Security within the database is modeled after a Roles and Permissions Matrix with the addition of pumps the user is authorized to access. When a user is defined to the system, a role is given to that user as well as authorization to pumps within the system. The role provides permissions for access to and manipulation of the pumps they are assigned. However, the system is also designed to allow the system administrator to modify specific permissions over specific pumps. For example, a field technician is authorized to add Human Incident Reports as well as Maintenance Reports, but is only allowed to view basic pump information. Suppose a field technician is a part of a community water committee. That technician may also be given access to the cost of ownership information for the pump within his community.

Current roles within the system include:

1. Administrator
2. Basic User
3. Donor
4. Field Technician

Data Triggers. A series of database triggers are executed with the initial insert of the un-parsed sensor data. As part of the parsing routine, the JSON string is validated and if it is determined that the JSON string is invalid, along with sending an ERR message to the monitor, the "bad data flag" is set in the Unparsed Sensor Data table. The insert of the un-parsed sensor data triggers a process to check the bad data flag and if it is true, insert data into the Historic Status table.

If the un-parsed data is successfully parsed, another trigger performs calculations on the sensor data and inserts the appropriate data into the Sensor Calculations table. That insert triggers yet another routine that derives status values based on the stored calculations and open HIRs and writes the data to the Historic Status table.

Periodically a script is run to check each pump for data in the Unparsed Sensor Data table. If no data is found for a configurable amount of days, the pump status is set to "grey" and data is entered into the No Sensor Log table. This triggers another routine which updates the status of the pump and inserts the data into the Historic Status table.

On any insert into the Historic Status table, another trigger is executed which determines if an alert needs to be sent. If so, the alert data is stored in the Pump Alert table and yet another trigger updates alerts for the appropriate users.

Triggers provide an automated method of identifying the status of pumps and sending the alerts to the appropriate users.

4.4   Web Application

The data from the monitor is collected and stored automatically in the database which can be accessed by authorized users via a secure web application or mobile app. The status of individual handpumps can be viewed on a map interface powered by Google Maps®. Each handpump on the map is represented by the pump status indicator (green, yellow, orange, red, or grey) depending on the level of functionality of the handpump.

The reporting module allows users to select a time period, single or multiple handpumps, or pump groups for further investigation, and provides either detailed or aggregated information. Selection can be accomplished via the map interface by selecting individual handpumps or pump groups. The IWP web application can export these queries as printable PDF reports or MS Excel Spreadsheets for further investigation or reporting purposes.

Notifications are handled automatically by the IWP software in the event of a change in a handpump status indicator. These are sent to appropriate stakeholders via email or text message depending on their preferences. The notifications are sent in the case of degradation, such as a status change from green to yellow or red, and in the case of an improvement, such as a status change from red to yellow or green. This allows stakeholders to know not only when a pump is broken, but also when and to what extent it has been repaired.

Technology Stack. The technology stack was decided upon with an effort to leverage open-source technology for the sake of Non-Profit Organizations who may not have large budgets. Therefore, the following was decided upon:

1. Server Software: Ubuntu version 12.0
2. Web Server Software: Apache version 2.0
3. PHP version 5.4.6
4. MySQL version 14.14
5. Smarty version 3.1
6. Bootstrap version 3.0.0
7. jQuery version 1.10.2

Application Structure. The Web Application was designed to abstract the presentation layer from the business logic layer and the business logic layer from the data layer. The presentation layer utilizes Smarty, a PHP Template Engine [10]. The data layer is encapsulated in PHP Classes. The business logic layer is written in php files within the public_html folder of the application. This decision was driven by the desire to insulate the application from database changes so that developers are unaffected by changes.

Reporting and Analytics. As mentioned in a previous section, the insertion of data into the database triggers a process that determines the current status of the handpump based on the current sensor data and human incident reports.

Fig. 4. Pumps on a Google Map®. (Color figure online)

A handpump will have the status of Green (running fine), Yellow (concerns exist about the handpump), Orange (significant problems exist with the handpump), Red (pump failure), or Grey (handpump has not been heard from). This status will be displayed on an authorized user's dashboard and the handpump location depicted on Google Maps® (See Fig. 4). If the status of the handpump changes, the alert system notifies the appropriate authorized users of the change. Handpump status is a configurable function defined for a given set of handpumps based on the values of daily volume extraction, leakage rate and maximum well level. An administrator has the authority to configure the Yellow and Orange statuses knowing that a pump more healthy than Yellow is Green and a pump less healthy than Orange is Red. Grey has been chosen for pumps from whom no sensor data has been received for a configurable number of days.

Google Maps® Integration. In order to make the web application user friendly, pumps were displayed as an overlay of a Google Map® utilizing the Google Maps® JavaScript API [11]. When a user logs in and selects the Pump Map, the general location of their pumps shows up on a Google Map® within the web application. The default zoom is applied and pumps are shown as circles in aggregated groups. As the user zooms in, the pump aggregation becomes individual pumps. The pump circles are colored based on the pump's status. Pumps may be selected from the map to view the Pump's profile.

On the Pump's profile page, graphs of the pump's leakage rate, longest prime, and volume extracted may be viewed along with the pump's basic data (Pump Name, Phone Number, Activation date, and GPS coordinates). The date range may be changed by the user to see specific days or ranges of days. Along with the graphs is a table of the detailed sensor data.

Graphs of multiple pumps at a time may be viewed by selecting multiple pumps on the map and choosing the aggregate view which will display the pump data together.

Along with the Google Maps® interface, we also support the ESRI system [12] by exporting a KML file using the Google Maps® API which in turn is imported into the ESRI system.

Some future mapping considerations are to allow other research organizations to overlay data with data found in IWP to get a better picture of what is happening in any given region.

4.5 Mobile App

A mobile app is an important element of the overall IWP system. It serves as a lightweight tool for workers to use while in the field, thereby enabling them to take advantage of mobile services such as location awareness and offline modes.

User Authentication: Mobile app users login to the mobile app using the same user name and password credentials of the web application. The mobile app authenticates users using the same web backend system that authenticates web users. Furthermore, the mobile app limits user access to handpump data and functionality identical to the security and access model imposed by the web application.

Mobile App Interface: Figure 5 displays the main screen of the mobile app. The core IWP mobile app functions are handpump initialization, filing maintenance reports, filing incident reports and viewing handpump alerts.

Handpump Initialization: This function allows a user to initialize a handpump into the IWP system. The function records the field technician assigned to the handpump, the GPS coordinates of the handpump (either automatically recorded (default) or entered by the user), the phone number on the SIM card installed in the handpump monitor, the date/time of the initialization and other descriptive information about the handpump.

Maintenance Reports: This function allows a user to create and submit handpump maintenance reports (as previously described), including the identity of the handpump, the user filing the maintenance report, date and time of the maintenance report, a brief description of the maintenance performed on the handpump and the total cost of the maintenance report broken down by travel, part, and labor costs.

Incident Reports: This function allows a user to create and submit handpump incident reports (as previously described), including the identity of the handpump, the user filing the incident report, date and time of the incident report, a brief description of the incident being reported on the handpump and the nature of the incident.

Fig. 5. IWP mobile app.

Alerts: This function allows a user to view the alerts (as previously described) associated with the handpumps for which the user has been granted access.

Offline and Synchronization Modes: The mobile app requires a data grade connection to transmit data, and since there will be times when such a connection is not available, an offline and synchronization mode is necessary. In such cases, the mobile app will locally persist the data (i.e. yet to be filed maintenance reports and incident reports) on the mobile device. Upon the mobile app sensing a data grade connection, the persisted data will be transmitted to the IWP system.

There are several important future considerations for the IWP mobile app. The current mobile app is purpose-built for the Android Operating System® and uses the Android SDK®. Given the trade-offs of platform-specific mobile apps versus responsive web app design, we would opt for a responsive web design in the future that supports a myriad of mobile device types.

Secondly, we would incorporate location awareness features in the app that would allow a user to check for pumps near them or to be given directions to a specific pump. We would also extend the app to include multi-language support. Finally, we believe pumps should become part of the "Internet of Things" ("IoT") economy and themselves become network addressable using NFC, thereby allowing authorized users to obtain data about a pump by using an NFC-enabled mobile app.

5 Summary and Future Work

A key purpose of this paper was to outline the key software design decisions made for IWP. The real-world environmental factors for deploying IWP (climate, lack of reliable Internet access, solar power, etc.) all made for a set of interesting design constraints. Hopefully, this paper provided some insight into the key software design decisions made and rationale for the same against these imposing design constraints.

Significant opportunities remain for advancing our work. Sensor data collected over time will validate assumptions and features such as alerts, change in handpump status, etc. Ideally, we would like to see the system deployed through the status cycle of a handpump to insure that the system correctly senses the deterioration of the handpump, issues the appropriate alerts and senses the handpump performance improving upon being repaired by a field technician.

In addition to the future directions of IWP outlined in the paper, the IWP software will continue to evolve based on lessons learned from field trials. It is expected that significant advances will be made in handpump data analytics based on feedback from handpump field technicians. Beyond the ideas mentioned in the paper, there are a number of future directions envisioned for IWP, including:

1. publishing and supporting APIs (data import, data export, big data analysis, etc.) and a platform for third party application developers to integrate with, innovate on and extend our platform and the capabilities provided by IWP,
2. creating a collaboration capability for researchers, field water technicans, community leaders, NGO staff and funders to be able to dialog and share information about pump performance and related data,
3. creating anonymous data aggregation and analyses capabilities for organizations and other interested parties to compare their pump performance and related data to such data from other pump groups.

Acknowledgements. We acknowledge our hardware counterparts, Dr. David Vader and Anthony Beers along with various Computer Science and Collaboratory students for their work on this project.

References

1. RWSN: Handpump data 2009 (2009)
2. Nejmeh, B., Weaver, D.S.: Leveraging scrum principles in collaborative, interdisciplinary service-learning project courses. In: Frontiers in Education Conference (FIE), pp. 1–6. IEEE (2014)
3. Weaver, D.S., Nejmeh, B., Vader, D., Beers, T.: The intelligent water project: bringing understanding to water pumps in Africa. In: Proceedings of the 2nd International Conference on Geographical Information Systems Theory, Applications and Management (GISTAM), vol. 1, pp. 211–218 (2016)
4. Beers, A.Q., Vader, D.T.: India mkii pump sustainability study report. Technical report, Messiah College (2013)

5. MacArthur, J.: Handpump standardisation in Sub-Saharan Africa (2015)
6. Sansom, K., Koestler, L.: African handpump market mapping study (2009)
7. Crockford, D.: The application/JSON media type for Javascript object notation (JSON)(2006). http://tools.ietf.org/html/rfc4627
8. Nejmeh, B.A., Dean, T.: The charms application suite: a community-based mobile data collection and alerting environment for HIV/AIDS orphan and vulnerable children in Zambia. Int. J. Comput. ICT Res. **46** (2010)
9. Upside Wireless (2016). http://www.upsidewireless.com/
10. New Digital Group, Inc.: Smarty template engine (2016). http://www.smarty.net/
11. Google: Google maps APIs (2016). https://developers.google.com/maps/document ation/javascript/
12. esri.com: About Esri (2016). http://www.esri.com/about-esri#who-we-are

Author Index

Printed in the United States
By Bookmasters